ONE SIGNAL
PUBLISHERS

ATRIA

A HISTORY OF THE WORLD IN SIX PLAGUES

How Contagion, Class,
and Captivity Shaped Us,
from Cholera to COVID-19

EDNA BONHOMME

ONE SIGNAL
PUBLISHERS

ATRIA

New York   Amsterdam/Antwerp   London   Toronto   Sydney/Melbourne   New Delhi

ONE SIGNAL
PUBLISHERS

ATRIA

An Imprint of Simon & Schuster, LLC
1230 Avenue of the Americas
New York, NY 10020

First One Signal Publishers/Atria Books hardcover edition March 2025

ONE SIGNAL PUBLISHERS / ATRIA BOOKS and colophon
are trademarks of Simon & Schuster, Inc.

Some names and identifying characteristics have been changed.

For information about special discounts for bulk purchases,
please contact Simon & Schuster Special Sales at 1-866-506-1949
or business@simonandschuster.com.

The Simon & Schuster Speakers Bureau can bring authors to your live event. For
more information or to book an event, contact the Simon & Schuster Speakers
Bureau at 1-866-248-3049 or visit our website at www.simonspeakers.com.

Interior design by Davina Mock-Maniscalco

Manufactured in the United States of America

1   3   5   7   9   10   8   6   4   2

Library of Congress Cataloging-in-Publication Data has been applied for.

ISBN 978-1-9821-9783-4
ISBN 978-1-9821-9785-8 (ebook)

Pou Manman, Grann, Melo, et Fifi

"Most men today cannot conceive of a freedom that does not involve somebody's slavery. They do not want equality because the thrill of their happiness comes from having things that others have not."

—W. E. B. Du Bois, *Darkwater: Voices from Within the Veil*

Contents

PROLOGUE   xi

CHAPTER 1: Contagion on the Plantation   1

CHAPTER 2: The African Laboratory   39

CHAPTER 3: Who's Afraid of the Flu?   73

CHAPTER 4: Breaking the Walls of Silence   105

CHAPTER 5: Ebola Town   139

CHAPTER 6: Relentless   183

CHAPTER 7: Locked Up   219

POSTSCRIPT   231

ACKNOWLEDGMENTS   235

ENDNOTES   239

INDEX   269

ABOUT THE AUTHOR   289

Prologue

Everyone who is born holds dual citizenship, in the kingdom of the well and in the kingdom of the sick. Although we all prefer to use only the good passport, sooner or later each of us is obliged, at least for a spell, to identify ourselves as citizens of that other place.

—Susan Sontag

The history of illness is not the history of medicine—it is the history of the world—and the history of having a body could well be the history of what is done to most of us in the interest of a few.

—Anne Boyer

RARELY ENJOY AN ENTIRE NIGHT of rest without interruption. I fixate on the whorls of brass of a light fixture, or a stray fly buzzing through the room. My mind lurches between fatigue and alarm. Like many insomniacs, my angst has an origin story— it began when I was a child and is bound in my memories like a cankerous sore.

In the summer of 1988, after I suffered nearly a week of high temperatures and perennial diarrhea, my parents—both in their early twenties—took me in desperation to Jackson Memorial Hospital, the same public infirmary in Miami where my mother would later work as a janitor. After a series of tests, I was diagnosed with typhoid fever. I spent a month there, at four years old, lonely and scared, curled up small in a hospital bed that felt huge. The stale air wafted through my nose while the EKG machine rattled through the night. I heard the buzzing of the fluorescent lights as the stiffness of the bed pressed up against my body. Unable to swallow without feeling nauseous, I could barely eat.

According to my aunts and parents, I nearly died. Hyperbole is a familiar register for my elders, their dramatic tales making for entertaining family gatherings, but for once their alarm at my sluggishness and flushed skin made sense. I was seriously ill. Unmoored from myself with a deep headache, my throat burned, and my cough echoed through the room. When my relatives visited me, I skulked along the ward corridors, dragging my tiny legs slowly from the bedroom into the playroom. But the worst part was that when the nurses arrived with my favorite fruit—mango—I barely touched the plate. Instead, I glared at them with repulsion, and at every other nutriment with irritation, as I crumbled into a state of fatigue. What had started as a fever turned into a curse that my miniature body, mind, and spirit could no longer bear.

My hospitalization stretched on, primarily to get me to a place of health, but also to ensure that I could not infect others. I lost weight. As each day turned into night, I heard the sounds of the nurses as they moved around shaken. My decline, my high temperature, and eventually my progress were left to the staff. Most of the time, they confined me to my bed and the room, unable to move as I pleased or join the games of my friends who I could see in the playground in my mind's eye. In my confinement, I felt like I was outside of myself.

This month of hospitalization has had a life of its own in the years afterward. In the early hours of the morning, as I am lying in bed, the early memory of captivity returns. I am reminded of the waves of tears that flowed down my round cheeks when my parents would leave and the sense of being trapped. When one is bedridden with a life-threatening disease (or a chronic illness), one might find oneself obsessed with death. I was too young to understand the cause or nature of my condition, but had a sense that death was something foreboding, a darkness hanging over me. Restless, I craved climbing the coconut tree in our backyard and searching for ladybugs with my cousins. My fear was driven by the conviction that I would not wake up the next time I closed my eyes. More than my parents' embraces and promises of safety muttered in Creole, or the icy ball of the nurse's thermometer, the hospital bed was the most consistent thing I felt.

The details of this period are less significant than the impact it created, how three decades later, I feel irrevocably out of place when entering a hospital. I absorbed the hushed voices, the pastel-colored walls, and the spartan hospital gowns. I could hear the doctors' voices but could not understand what they were saying. The syllables rattled through the air, meaningless sounds. Barely able to articulate my thoughts in either Creole or English, I was a callow girl lost in a room of tubes and bright lights, obsessed with my plans to escape. I couldn't see past my hospital bedroom, my mom, and the sense that my restraint was my punishment for pushing a friend off the slide, as I recalled doing in my local playground shortly before I was taken ill. All I could do was count the number of times my parents came per day, or the moments the nurses came to feed me. I had no control over anything—what I ate, where I went, when I moved. Even when I imagined playing with my friends at the beach or pictured the sugarcane stalks in our backyard, my thoughts were dimmed by the atmosphere. When others entered the room, even my family, I felt they were surveying me, taking the measure of the bed, the equipment, and my body.

Now, I find that a bed, even when lying next to a friend or lover, is also a place of unease. Will they disappear in the middle of the night once I am asleep, or will their presence comfort me? For months after my infection with typhoid fever, I felt the profound injustice of being forced indoors and restrained, unable to feel the humid, heavy Miami wind blowing on my face, or watch the luxury sailboats with sunburned passengers splash across the calm water of Biscayne Bay. I learned how to sit still with my thoughts. Even today, hospitals agitate me, not because of their stiff chairs or drab walls, or even their bland art, but because of my dread. Medical institutions remind me of my most profound envy—for people who can sleep a whole night without interruption or harm.

In the late nineteenth century, typhoid fever was one of the most prevalent diseases in America. When Mary Mallon, popularly known as "Typhoid Mary," was identified as an asymptomatic carrier of the disease, physicians believed that she infected families in New York City for whom she cooked by contaminating their drinking water with the bacterium. It was the same fear of contagion through food or water that led my doctors at Jackson Memorial Hospital to conclude I should be isolated from my family. Treatment was not just a matter of medication; it meant alienation from my parents, my friends, and my community.

My infection with *Salmonella typhi*—the bacteria that causes typhoid fever—was unusual, given the time and place. Eighty years before, typhoid fever was the fourth largest cause of death in the United States with one in every 1,000 deaths attributed to the disease. Between 1985 and 1994, there were 2,445 cases of typhoid fever in the United States. Eighty percent of them occurred in six states, Florida among them. Most of the cases were from people who had traveled abroad, but by that point, the farthest I had traveled was the county next to Miami. The cause of my infection was not immediately obvious, but what mattered was that I had access to antibiotics. Without treatment, my chance of death could have been as high as 20 percent. Currently, typhoid fever is endemic in Asia, Africa, Latin America, and the Caribbean.

"Most of the unpleasure that we experience is *perpetual* unpleasure," Freud once wrote of anxiety. "It may be a perception of pressure by unsatisfied instincts; or it may be external perception which is either distressing in itself or which excites unpleasurable expectations in the mental apparatus—that is, which is recognized by it as a 'danger.'" For me, confinement is the device that evokes danger. I have learned to take hold of this feeling, to grasp and try to control it, by peering into the ways my family, and others, have been perceived to be ill and as a result have been confined against their will. From state prisons to mental health institutions, confinement is a ubiquitous feature of society, a crucial buttress to the systems of power. Forced captivity following an outbreak of disease is partly about blaming a particular person or group of people. Captivity is political, and it maps on to the many ways we already deliberately segregate society. My hospitalization at Jackson Memorial was not the only time that confinement occurred in my family; a similar phenomenon happened in our community several years before.

We Are Our Ancestors

In 1979, my father took a boat across the Caribbean Sea from the coastal town of La Baie in Northern Haiti, leaving his family behind. Like most Haitian peasants, they lived on the land that had been redistributed to African-descended people who liberated themselves in 1804. Their farm, a symbol of collective emancipation, had been passed down from generation to generation, but after two centuries of intensive agricultural production that had degraded the soil, years-long drought, and a series of hurricanes, my father and his relatives could no longer live off the depleted soil. My grandparents, their parents, and their community of subsistence farmers and fishermen struggled to feed themselves. The coastal village of my father's birth sits on the Caribbean Sea, with no paved roads or running water. To this day, La Baie lacks electricity and running water. The whistling tune of the wind and the roar of the ocean were far more familiar to my father during his childhood than honking cars.

As the late Haitian anthropologist Michel-Rolph Trouillot wrote in his book, *Haiti: State Against Nation*, "While the state turned inward to consolidate its control, the urban elites who gravitated around that state pushed the rural majority into the margins of political life." In some cases, they were pushed out of the country. The dictator Jean-Claude Duvalier, with the cooperation of upper-class Haitians, subdued dissidents and the impoverished through intimidation and violence. My father, mother, grandparents, aunts, and uncles were opponents of the Duvalier regime. But political repression wasn't the only problem. With only a high school education and few prospects for work in rural Haiti, he was in every sense excluded. Political oppression and abject poverty were entwined with each other as motives for his escape.

He was not alone. Between 1972 and 1981, between 55,000 and 100,000 Haitians migrated to southern Florida by sea. Some boats were barely seaworthy, others would capsize, drowning their passengers in the choppy waters of the Caribbean Sea. My father's journey took him on an overcrowded vessel, briefly passing through the Bahamas before embarking in the lower Florida Keys. He survived the journey, and a brief period of detention by the US Immigration and Naturalization Services (now called the U.S. Citizenship and Immigration Services), but quickly faced the new challenge of finding a livelihood and safety in his new home. Without training or skills, and speaking only Haitian Creole and French at the time, his options were limited. Nevertheless, he was intrepid. He joined other working poor Haitians in Lemon City, Miami, doing manual tasks as a day laborer. Eventually he became a salaried sanitation worker at a factory.

Two years later, my mother would make the same boat journey. By then, Lemon City had taken on the informal name "Little Haiti." Situated fifteen minutes north of downtown Miami, the neighborhood was saturated with colorful homes, decorated with figures of Catholic saints, and surrounded by banyan fig trees, stray chickens, and men playing dominoes in their yards. My neighborhood was home to Veye-Yo, a Haitian language radio station where immigrants shared music, gossip, and their visions for the political future of Haiti, forging a new reality in their adopted land. Little Haiti echoed the traffic of Port-au-Prince and the briny aromas from Jacmel. But to my parents' dismay, by the early 1980s, the community in which they found safety in Miami was already stigmatized and feared, blamed as the source of a novel epidemic—HIV/AIDS. The meaning of the neighborhood was transformed; no longer a haven for the community, but a zone of confinement.

When HIV/AIDS emerged as a modern-day plague in major cities in the United States in 1981, people of Haitian descent were perceived to be a risk group. As early as 1982, the United States Centers for Disease Control and Prevention listed four supposed groups as a "high risk" for contracting or transmitting HIV/AIDS: homosexuals, heroin users, hemophiliacs, and Haitians. Haitians were the only members of the "4-H club" included on the basis of their nationality. As a consequence they and their US-born children were denied housing and employment. Working-class Haitians found themselves marginalized four times over: They were Black and so belonged presumptively to the US underclass; they were poor; they were migrants; they were marked out as diseased. For many, Little

Haiti became their home by default: They could not rent an apartment elsewhere. Their containment—like many other such responses to epidemic diseases in history—was unequal in practice, even if not always in intent. The Haitian community of Miami, and by extension the United States at large, bore the unwarranted and outsized burden of scapegoating. For escapees from a dictatorship, this state of dismissal was all too familiar.

Trauma eats the soul and erodes memories. When I ask my parents about their migration from Haiti, their narratives contain gaps. They won't always name historical figures or events directly, and they don't mention François and Jean-Claude Duvalier, the dictators who ruled from 1957 to 1986, or the Tonton Macoute, the secret police that terrorized dissenters. They don't mention the friends they made or the friends they lost. They tell stories of the mundane and familiar experiences of making plantains, speaking Haitian Creole, and passing on the lessons of bodily care. Like the dictatorship, slavery and colonialism cast a shadow over their testimonies, but it is too faint to grasp, too prodigious to name.

What matters more, in effect, is their perseverance. Being Black and free has been at the core of their being precisely because our ancestors—participants in the Haitian Revolution—actively fought for their emancipation. Black freedom means rejecting the lies that are often told and the misconceptions that are often believed about people who look like us—that we're lazy, quixotic, and bovine. Black liberation means knowing that we are worthy of dignity, respect, and life. This legacy, of course, is not unique to Haitians—it's one we share with other descendants of African slaves in the Americas.

Black survival—post–chattel slavery—is not just a matter of forgetting. There is a commitment to remembrance. In her book *Dear Science*, the Black scholar Katherine McKittrick explains the necessity of bards—within the African diaspora—as a way to transcend historical trauma. "We tell and feel stories (in our hearts), and this telling-feeling tells-feels the empires of black life." Like McKittrick, I seek to narrate the past by sifting through the ambiguities and contradictions to excavate, with patience and diligence, the subtle ways that chronicles defied our historical wounds. I include my family's narrative—for all that it is incomplete—to illustrate how contagious outbreaks produce hierarchies of life. In Little Haiti, whether on the mild winter days or the balmy summer nights, I felt held by a neighborhood that incubated my precocious mind with the whistling tune of *kompa* and the bright faces that glowed with contentment to see the children frolic in the streets. For years, Little Haiti, a community subject to confinement, was sprinkled with luster like the stars against the night sky. Most people were attentive and thoughtful and steadfast in watching over each other. I found my home among the waves of migrants pouring every bit of their souls into one deliberate act: survival.

Haitian migrants, like my parents, made their own rhythms and cadences in our neighborhood. They worked whenever and wherever they could—as seasonal domestic workers in Miami's beachside hotels, or cleaning bodily waste from the city's hospitals, or on hotel construction sites on Miami Beach. They partied on Saturdays and prayed on Sundays. Some of my family members, like my aunt Riri, stood on the corner of our block with other undocumented migrants, waiting for a yellow school bus that took them to work on a farm in Homestead, where they would pick beans and tomatoes. Haitians in my community struggled and loved, even if from the outside, they were scapegoated and loathed.

Being perceived as a vector of HIV was not the only thing that marginalized Haitians. Haitian Creole was a social fortification, but it instantly set us apart from the other migrants in Miami, most of whom spoke Caribbean variants of English or Spanish—especially the latter. Indeed, approximately 77 percent of the city's residents speak a language other than English, and Spanish is by far the most common. We were misunderstood. As I grew older, I learned to transpose myself between two worlds, never truly realizing what it meant to be both in place—within Little Haiti—and out of place in the rest of Miami. It's not just that we adapted to a situation we did not create, but that other people's anxieties and prejudices about Haiti and Haitians left us excluded, a social and medical malady.

In general, Afro-Caribbean migrants in the United States suffered the double burden of being both Black and migrant, with the working poor, like my parents, confined to a few neighborhoods. Yet we Haitians were hardly the first or last group of people to be blamed or stigmatized over an epidemic. This book is a journey toward understanding how disease management is influenced by how society defines humanity.

Humans and microbes live in constant communion, in a relationship that is at once antagonistic and beneficial. This delicate liaison is altered if the microorganism is associated with anxieties about sex, ethnicity, or the risk of death. Microbes themselves are blind. They know nothing but the need to reproduce, in whatever environment they find themselves. But the life cycle of the diseases they cause rests on the political order of things: the medical advancements that come with understanding the science of contagion, as well as the social dynamics that form our environment.

Medical racism, and the histories of how race and class shape the medical field, are not entirely new subjects for medical historians and science journalists. Black scholars in particular have explored themes such as premature death, as in Dorothy Roberts's *Killing the Black Body*, or the separate and unequal medical system in Harriet A. Washington's *Medical Apartheid*. In her book, Roberts exposes how the US legal system was discriminatory and negatively impacted African American health. Similarly, Washington offers a historical framework for conceptualizing medical racism. For Washington, "trying to ameliorate African American health without understanding the pertinent history of medical care is like trying to treat a patient without eliciting a thorough medical history: a hazardous, and probably futile, approach." Racism may not operate in identical ways or produce identical outcomes in every context, but there is an abundance of evidence, across many times and places, for its health effects. Tracing this history is so important precisely so that society can undo these health inequities.

Identity, in itself, does not guarantee that one will become infected with a virus. As the science journalist Linda Villarosa puts it in *Under the Skin*:

> The *something* that is making Black Americans sicker is not race per se, or the lack of money, education, information, or access to health services that can be tied to being Black in America. It is also not genes or something inherently wrong or inferior about the Black body. The *something* is racism.

Health and sickness are not just matters of individual choice or the physiology of individual bodies, but a reflection of a society's history—and its present—and the way its values are embodied in social structures.

In general, health is worsened by physical and mental stress, but a corporeal demise not solely come from within. "Illness is not simply caused by a foreign entity," physician Rupa Marya and journalist Raj Patel write in their book *Inflamed*, "it is the body's response to damage that may or may not be precipitated by that entity." Marya and Patel show how the many toxins to which the contemporary person is exposed, such as microplastics that cause infertility or pesticides that shorten our lives, injure our bodies in the name of a capitalist system that insists only upon the right to accumulate profit. In a 2022 study in Italy, microplastics were detected in the breast milk of three-quarters of the women who had recently given birth. Most of us cannot escape these effects unless we are part of the ultra-elite. The expensive fitness membership plan cannot erase the exposure to plastics. The wheatgrass spirulina "juice" may not add extra years to your life. Even the medical community, Marya and Patel write, may not save us:

> Most doctors—most humans, really—have unwittingly inherited a colonial worldview that emphasizes individual health, disconnecting illness from its social and historical contexts and obscuring our place in the web of life that makes us who we are.

Many factors shape our physical and mental health: where we are born, our social class, educational history, gender, genetic issues, and physical environment. The United States is a particularly harsh place to grow up—life expectancy has fallen by three years since 2020—mostly due to Covid-19, but also the worsening of people's quality of life. Researchers found that the effect could not only be attributed to the pandemic, but the other factors also included drug overdoses, accidental injuries, and suicide. Many features of the United States that have driven this fall in life expectancies relate to self-harm and coping with a society that provides very little to its citizens, which is unprecedented for a rich nation outside of wartime, including a health system driven by private profit, the closure of public hospitals, the busting of unions that fought for better wages and healthcare plans. These phenomena have manifested themselves in part in a relentless assault on racialized people, particularly those among the impoverished and working class. Their ability to eat well, to take sick days, to rest, to enjoy clean air, to have adequate housing, to access effective preventative medicine—all are diminished. Constrained by poverty, they are more likely to become sick. Captivity and contagion have always been intertwined.

Segregated spaces might be perceived as petri dishes of disease, but they also function as sites of resistance. This book sheds light on how people make space to heal in these medical and political enclaves of necessity. Although places like Little Haiti were scapegoated, as my parents found, they were also an oasis for refugees and their children to speak Creole, share art, and be soothed by the subtropical breeze. Epidemics have been shaped by the history of forced captivity—one that began on the plantation, in medical experimental camps, racial apartheid, and continues in immigration detention centers and prisons. But in each case, those held captive have resisted. They have always made the choice to be free.

In the hypnotic rhythm of infestation, as people worked out the details between preventative and palliative treatment, captivity has at times been a central feature of survival and punishment. This is the main subject of *A History of the World in Six Plagues*. **The main argument of this book is that pandemics start small, grow large due to negligence, and leave rot behind that we generally don't bother to clean up before the next pandemic arrives. I find that humans seek easy answers in crisis, but in so doing lay the groundwork for far more profound problems later.** As a historian of science trained in biology and public health, I analyze those histories with acuity, and as a working-class person of Haitian descent, I approach pandemics with compassion. Our personal identities aren't enough when it comes to understanding and addressing these issues, but neither is our scholarship, which is why we need both.

Captive Contagions

We all struggle with some form of captivity at some point in our lives. Of course, for a newborn unable to walk and a prisoner of war, confinement looks very different. Some of us find comfort in closed quarters: shielding ourselves from potential predators or protecting ourselves from a natural disaster. Others want nothing more than to escape. Captivity offers a useful prism through which to understand the various actors and agents that shape the paths diseases take, because it can be both a function of illness itself— the loss of freedom that attends serious infection—and part of a state's response to a new epidemic. Epidemics reveal the true nature of our political commitment to the notion of freedom. If, as the scholar Orlando Patterson noted in his seminal work *Freedom,* "freedom was generated from the experience of slavery," then the people who are made captive might be the most invested in a desire for liberation. This might be true at the level of personal autonomy, but also for entire communities.

Whether it is the right to obtain a safe abortion or the ability to access cancer treatment, working-class people and racial minorities face an array of challenges in securing the health services they need. Their struggles have only become more pronounced in the age of Covid-19. It has been devastating, to say the least, to witness in this pandemic the stark inequalities of both the US healthcare system and—in the form of the global vaccination rollout—healthcare around the world. Access to testing, treatment, and protective equipment have been inadequate for so many people, especially essential workers, who in the Global North are disproportionately people of color. The horrors are predictable. Even as we face new epidemics, each plays out an ultimately familiar tale of racial health inequalities, from reproductive health to environmental racism.

Captivity governs how bacteria and viruses coexist with us. The diseases they trigger reflect the social divisions that push some people to the periphery of society: unhoused people, sexual minorities, and racialized people. Some seem intent on forgetting the lessons of history, while others are forever haunted by the pain they cause, the memories that linger. In this way, epidemics help us understand inequalities, and history guides us in understanding the ways that health inequalities operate in pandemics today.

Where We Divide

Early on in the Covid-19 pandemic, the novelist Arundhati Roy wrote:

> The tragedy is immediate, real, epic, and unfolding before our eyes. But it isn't new. It is the wreckage of a train that has been careening down the track for years. Who doesn't remember the videos of "patient dumping"—sick people, still in their hospital gowns, butt naked, being surreptitiously dumped on street corners? Hospital doors have too often been closed to the less fortunate citizens of the US. It hasn't mattered how sick they've been, or how much they've suffered.

Inadequate and unequal healthcare systems long predate Covid-19. Almost everywhere, to be poorer means to be at greater risk, if one has not already fallen out of the healthcare system altogether and into despair. Even the necessary and laudable act of exercising caution out of sympathy for the immunocompromised felt like a choice that was only available to those who had the power and privilege to remain safe. For many ethnic and racial minorities in the Global North—who, like my relatives, did the invisible and dirty work of society—cleaning and cooking and delivering goods—transmission and death rates were disproportionately high.

More than thirty years after my hospitalization in Miami, I live five thousand miles away in Berlin, Germany, where I have made my home. When Covid-19 emerged, I felt a fresh wave of anxiety surface. As countries across the globe went under lockdown, some expressed great hostility toward the restrictions, denying the existence of Covid-19. Others pushed for "Covid Zero" policies, advocating for drastic isolation measures. This was not simply a case of isolating the four-year-old me in a hospital ward, or my parents in Little Haiti—most of the world reckoned with how to contain the contagion. Many of those who survived were left confused, angry, out of work, or suffering long-term health effects. Conspiracy theories floated over social media; anti-Asian discrimination ran rampant, even in Berlin. Activists created mutual aid groups, living situations changed—I moved in with my partner, and have lived with him ever since—and life, as we once knew it, completely changed.

Why did the lockdown leave us so jaded, and how did it manifest in ways that saw humans focus their fears on other people more than the microbes themselves? To understand this, I dug deep into archives, plantation manuals, ad hoc documentaries, interviews, and the work of historians both long-gone and active today.

The effects of a bacterial or viral infection on any given body can be understood at a biological level, and also by how people interpret a disease. Nevertheless, it is a challenge to understand our relationship to a microbe when we are all suffering the aftermath of an epidemic. In practice, it is difficult to come up with clear answers on how to cope with an illness, let alone care for each other. The response of some people to this challenge is to simply adopt a vacuous language of "freedom," which is often considerably removed from a practice of care—something many have noticed with dangerous, QAnon-type fringe groups.

Early on, during the Covid-19 pandemic, one of my friends was skeptical about the severity of the disease, often citing obscure scientists and dubious data. In discussions with him, I tried to think deeply about his understanding of the disease and work through the growing tension in our friendship. We found we could at least agree on one core value, the importance of bodily autonomy, even if we disagreed on how to achieve it through evidence-based public health policy. This book makes a case for personal sovereignty, but it also recognizes the importance of having solidarity with marginalized people in society. The vital task is to recognize that while these values may sometimes appear to be in conflict, they need not be—but only when we have built a whole society based on an ethic of care and respect for all. Ultimately, these are stories of personal survival, and struggles for safety and health, and against death. They are not primarily a tale of scientists who discovered illnesses or devised new treatments; I try to sit with individuals and communities struggling against disease and confinement and fighting for their humanity.

"Illness is the night-side life," Susan Sontag remarked, "a more onerous citizenship." My childhood sickbed will always stay with me, a constant reminder of a confinement that came to symbolize every barrier I subsequently encountered. I recognize versions of myself in the histories that I present here, the people haunted by perennial illness, like Virginia Woolf, or those who are confined to their neighborhood by military occupation, such as residents in West Point, Liberia. I see myself in the activists striving to destigmatize a modern plague and working to find solace in the community, in the resilience of the grassroots political group AIDS Coalition to Unleash Power.

A History of the World in Six Plagues investigates the various lives of diseases, from their development in the infected person's body through the ways that social or racial categories influence the path of infection, to the large-scale policy responses of states and governments. The book begins with a significant nineteenth-century pandemic—cholera—exploring the international response and its relationships to freedom. I examine how people move in and out of states of illness, and at different times and with different motives have demanded or refused confinement. These social questions are not ancillary to the study of disease but integral to it.

These epidemics are not just isolated case studies; together they tell a history of the world. Studying the plantations, concentration camps, sickbeds, prisons, slums, and private homes brings us to a new understanding of the ways confinement tells us about the world. From cholera to Covid-19, it is impossible to understand who gets sick without thinking about how confinement works. The outbreaks in the early chapters occurred before the development of antibiotics and mass immunization campaigns, with a focus on cholera and maternal mortality on the plantation, sleeping sickness in colonial concentration camps, and the effect of influenza on those who have been bedridden. The second part of the book focuses on the present—in a new era of illness where mass data collection and surveillance have occurred—HIV/AIDS in a prison, Ebola in an African neighborhood, and Covid-19 accounts in the first two years of the pandemic.

Such is the scope of the changes now taking place that I could have gone pretty much anywhere and, with the proper guidance, found signs of them. Understanding these processes requires that we pay special attention to the lasting effects of chattel slavery, colonialism, war, and incarceration—not as bygone abstractions but as perennial marks that continue to shape our health.

Anyone who has lived through the Covid-19 pandemic, let alone those who lost a relative, knows what it is to feel conflicted over the question of confinement. On the one hand, it can be important to separate people when there is a communicable disease outbreak. But on the other hand, social isolation possesses grave danger faced by those who are perennially confined in crowded houses and workplaces. This is an account of the many ways contagion has made people feel disenchanted and dispossessed, but it also explores how captivity might create new communities—especially when people need to form social networks to survive under duress. This book tells the accounts of people who deserved better. It is also a story of redemption, and of the little child in all of us, curled up alone in a huge bed, without her parents, who wants to be healthy and free.

Chapter 1

CONTAGION ON THE PLANTATION

He who has health, has hope. And he who has hope, has everything.

—Proverb

The hold of slavery was what I sought to articulate and convey. The category crisis of human flesh and sentient commodity defined the existence of the enslaved and this predicament of value and fungibility would shadow their descendants, the blackened and the dispossessed.

—Saidiya Hartman

ON A COOL AUTUMN DAY in New Orleans, Dr. Samuel A. Cartwright stood in front of a pulpit, facing a couple dozen learned men. He was an insolent man with an owl-like face, slightly balding, a zaftig body that commanded the room. That day, the air was crisp, a reprieve from southern heat. The sixty-four-year-old man had gained popularity from other southern gentlemen. It's not certain if he spoke with a staccato rhythm of a preacher or if it was the low hum of a bass singer bellowing with an infectious tempo, but his duties were clear: to deliver a lecture on his theories on the anatomy and physiology about "the negro" race.

On November 30, 1857, several years before the US Civil War, Dr. Samuel A. Cartwright delivered a lecture at the New Orleans Academy of Sciences. As the city of New Orleans grew from 10,000 people in the beginning of the nineteenth century to 100,000 residents in the mid-nineteenth century, the institute was a haven for southern men who wanted to share their expertise in the natural and social sciences. Founded in 1853 by several white physicians, the members were experts in geology, anatomy, and medicine. In some cases, the materials were concerned with structural projects, such as protecting New Orleans from flooding, but in other instances, the professionals who addressed the academy did so to champion the American South's most lucrative institution: slavery.

During his hour-long speech, Dr. Cartwright offered his theories about the physical differences between the audience and what he referred to as the "prognathous race." The observers were all white, and the "prognathous race" in question was Black. As a leading physician from Mississippi, Cartwright's essays, often published in southern medical journals, were replete with rhetoric that claimed there were scientific differences between white and Black people, with the purpose to exculpate chattel slavery. During his speech to his colleagues, Cartwright alleged that Black people in the United States were "healthier, happier and more prolific than in their native Africa—producing, under the white man's will, a great variety of agricultural products, besides upwards of three millions [sic] of bales of cotton, and three hundred thousand hogsheads of sugar." He was right to point out the economic value of the enslaved, and their unpaid labor provided the United States with an economic boom, but these people were neither healthy, happy, nor prolific, especially given their captivity. This, of course, mattered little to Cartwright or his audience, their version of science fitting neatly into the hierarchies they wanted to maintain in southern US society.

Black Americans' subjection to confinement is a story about American physicians' role in upholding chattel slavery. It is the story about a society riven by the need to keep Black individuals infirmed, and a tale of willful white American ignorance, that thousands of enslaved people, even with fleeting failure, tried to avoid life on the plantation.

Cartwright argued that people of African descent were "primitive" and that the "subordination of the inferior race to the superior is a normal, and not a forced condition." Given that enslaved Black people were resisting and escaping from plantations throughout the South, Cartwright and his advocates needed a principle to explain why they had a "right" to confine a person in the first place. His reading of biology gave him an inaccurate but favored opinion for many white men who lived in the South. In 1857, near the tail end of US chattel slavery, four million Black people were living in the United States; 90 percent of them were in the American South, and most of them worked against their will on a plantation. As slavery grew, plantation owners depended on the "expertise" of physicians like Dr. Cartwright to counsel them about the health of the enslaved.

For the first half of the nineteenth century, white American enslavers relied on the prescription of physicians to diagnose and treat the enslaved, often leaning on the advice of practicing physicians who used science to justify Black bondage. In his 1832 book, *Some Account of the Asiatic Cholera,* Dr. Cartwright provided recommendations on how to manage the overall health of cholera victims, including enslaved people living on a plantation. While he attested that preventing cholera meant that slave cabins "should be aired and kept clean," there was very little action—by enslavers—to put this into practice. For the latter, unsanitary conditions on the plantation meant that disease transmission was rampant.

Despite the opinion of Dr. Cartwright, the slave cabin, both suffocating and densely packed, made it more likely for its inhabitants to be infected with a respiratory disease or recovering from an injury. Given that many lacked access to clean water and nutritious food, their life circumstances were designed for them to feel defeated. How can we understand why Dr. Cartwright's theories were so popular, even if the circumstances of their lives, not their biology, was the issue that made their existence so perilous? The contradictions of a doctor claiming to provide healthcare to Black people while knowingly expressing disdain for them show how insidious the southern medical institution was. Nevertheless, this was the beginning of modern medicine as we know it. When physicians began to navigate a biological terrain through invasive experiments against the will of marginalized people.

For most of the nineteenth century, the concept of a bacterium or a virus was not part of the medical landscape. Among the enslaved people who lived on a plantation in the American South, there was confusion and fear about the mostly white medical community, particularly because of the alliance they formed with their enslavers. To be a slave in the early nineteenth century was to be damned to cruelties above and beyond even the terrors of servitude. If a person was sick with pneumonia, that individual might not have time to rest. If she was giving birth, she might not have received pain relief. At best, an enslaved person would be given an arsenic compound meant to expel that bilious fever as was the recommendation by southern physicians at the time.

Dr. Cartwright, an anti-contagionist, rejected the claim that diseases were passed from one to another and assumed that illness emerged solely from the environment. He surmised that "whatever may be the cause of Cholera, its history informs us that a damp, confined and impure air gives efficacy to its cause; and want exposure, in temperature, fevers, terror or whatever disturbs the balance of the circulation, gives it subjects." Given that he assumed cholera emerged from lousy air, his solution to an outbreak entailed removing people from an enclosed space. He later added that if cholera surfaces "on plantations among the negroes, it will be necessary to consider the propriety of removing or scattering them, in order to prevent further attacks, and to mitigate its violence." Yet this suggestion was rarely implemented.

Another area where the enslaved were vulnerable was their reproductive health. After the United States discontinued importing enslaved Africans in 1808, the primary way that an enslaver produced more bondspeople was through forced reproduction. For most enslavers, Black women—who were often denied a choice about pregnancy—were responsible for birthing a new generation of slaves. The sexual assault of Black women and their lack of access to prenatal care meant they had little autonomy over their bodies. The problem fell outside of whether they became parents; it also meant they were privy to nonconsensual research.

Whether or not they called it an experiment, the act of being an enslaved woman meant that their owners could, without their consent, carry out grim and futile treatments. One planter's medical companion suggested that pregnant women undergo bleeding with the idea that the body "may be freed of an imaginary redundancy, not recollecting that the process of pregnancy is going on, to employ the interrupted menstrual fluid, to the now essential purpose of affording growth, and support to the increasing womb itself as well as its contents." These words are striking, precisely because they suggest that dehydrating the pregnant person would lead to a successful birth. In reality, anyone who has been pregnant would know that this would be an excruciating experience, one that would further lead a person to exhaustion. This suggestion was not the only treatment for pregnant women. Thomas Jefferson, in contrast, asserted, "Never bleed a Negro." This was on the basis that he surmised that treatment for Blacks and whites should be different, a view that Dr. Samuel A. Cartwright also argued. He noted that "the same medical treatment which would benefit or cure a white man would often cure or injure a negro." The inconsistencies between these nineteenth-century men are not unique. In fact, contemporary scientists often have conflicting viewpoints; however, the basis for Dr. Cartwright and Jefferson's premise, and ultimately their conclusion, lacks an evidence-based approach or sound logic.

In the early 1840s, an enslaved woman had pain germinating from her abdomen, having experienced constipation for several months. When William Patterson, her enslaver, and resident of Bryan County, Georgia, asked for the assistance of Dr. Stephen Harris, he prescribed the woman five grams of iodide of potassium. Several days later, the woman died. Her passing, in itself, was a tragedy, but what proceeded was that Dr. Harris conducted an autopsy on her reproductive organs, not by asking for the permission from her family, but from Mr. Patterson, the man who claimed ownership over her body. Overall, very little was done to ensure Black women's comfort and viability when they were alive, or their privacy when they were deceased.

Woven in a patchwork of remedial recipes executed in an unequal system, "plantation medicine" was the rubric southern physicians and enslavers embraced to sustain captivity. Those, like Cartwright, who advocated for "plantation medicine" undermined the health of all Black people who were held captive, even if the medical policies were seemingly "rational" for the time. Then and in practice, their actions and reagents were, at best, ineffective and, in some cases, outright hazardous. For example, Cartwright's recommendation for treating cholera was "an emetic of Ipecac, with a small bloodletting." What we know about ipecac today is that it induces vomiting and bloodletting has been largely abandoned by most Western physicians given its potential harm to patients. The lurid mythos that Black people were not meant to be free—on account of white enslavers and doctors—meant that when Black people were subjected to outbreaks such as cholera or susceptibility to maternal mortality—they had little to no treatment to cure their ills.

If the enslaved appeared physically weak or emotionally drained, it was not because of some "inherent" biological circumstance that was attributed to their skin color—instead, it had to do with the condition of confinement that made them more vulnerable to disability and mortality. Enslavement meant that they were more likely to be exposed to occupational injuries such as heat exhaustion and physical impairment. They were often overworked and underfed.

In developing a medical practice intended for captive people, contemporary social scientists argue that there is little denying how the state of the enslaved fueled mental and physical maladies. Historian and political theorist Achille Mbembe evinces that Western democracies facilitated the technologies of colonialism that denigrated the body through brutality, a mortifying reality wherein the everyday violence of chattel slavery created the conditions of making the enslaved more susceptible to illness. In their book *Cabin, Quarter, Plantation*, Clifton Ellis and Rebecca Ginsburg assert that enslaved people exercised power by learning the inner workings of the southern landscape and subsequently relying on their knowledge of the woods, fields, and trails to support their escape. They highlight how spaces of confinement, which were framed by their enslavement, and them being stowed away in an interior space, mentally and physically compromised enslaved people.

Plantation medicine's focus was neither to help nor to cure the enslaved. Overall, the medical manuals published by physicians such as Cartwright often provided detailed accounts of their crude experiments and baseless theories. Plantation manuals served as a guide for enslavers and working-class white people who lacked access to a hospital. There was no monopoly on a single medical plan for the enslaved, but what is clear from these texts is that slaves—especially Black women—had little input about the treatment they would receive. For their enslavers and the medical men they employed, they conjured their version of race science.

Plantation medicine was not neutral; it was a way to reify differences. Enslavers even took it upon themselves at times to read medical literature rather than get a professional's direct help. Reminiscent of recent history, where some U.S.-based Americans championed self-administering ivermectin for Covid-19, some enslavers believed castor oil and bloodletting could cure anything. As historian Rana Hogarth noted, enslavers invested in plantation medical science and a constellation of ad hoc medical practices, which "became an essential component to the development of the medical profession in the Americas," leading to what she refers to as the medicalization of Blackness. Hogarth argues that these theories served as a basis for widespread claims of biological distinction between the races, which, in some cases, hold space today.

Say nothing of the dread of being ill while tending to a tobacco field, sitting in a makeshift and poorly insulated shed after a day's work. Of the physical ailments that bud with other enslaved folk in the cabin. Of the fragile bond between the body and mind and not knowing when and how this would all end. The gap between the living condition of the enslaved and the alleged promise to cure by the plantation medical manuals was pronounced when enslaved Black women had premature births or the enslaved were poisoned by their masters with reagents that were alleged to treat them. Plantation medicine was a means to further establish a social hierarchy, under the guise of science. Doctors had the power to subside a disease or prevent death, when infectious diseases ravaged a plantation.

Cholera on the Plantation

At the beginning of the nineteenth century, nearly 40 percent of white families in Georgia owned slaves. Most of their captives were African American, who lived in the highly fertile land in the hinterland of the state, and the remainder of them lived on plantations along the Atlantic coast. Their geographical position wasn't what mattered, rather it was their economic value. Given that over 400,000 people (44 percent of the state's population) were enslaved, the state was deeply wedded to maintaining slavery. In material terms, enslaved people contributed more than $400 million to the state's economy—without being compensated. By the eve of the US Civil War, slavery accounted for half of the state's wealth. The economic significance of slavery to Georgia's economy and, in some cases, the death of bondspeople, especially during an outbreak, compelled some enslavers to intervene and try to improve their health. Some enslavers left tills about the costs, loss, and profit margins of their slaves; however, some chose to provide social spectrums of their lives. Louis Manigault, a Georgia man who owned several plantations, kept extensive records about his plantations.

Adjacent to the Savannah River, the Gowrie Plantation had, on average, one hundred enslaved people living on the grounds. Most of the Black captives resided in a cabin which contained at least four people in an eighteen-by-eighteen-foot single room. Often, these people were unrelated; in some cases, they could be moved to another one of Louis Manigault's plantations. The Gowrie Plantation, located in southern Georgia, was one of the most profitable

9

plantations of the region, precisely because of the brutality of enslavement. Dozens of Black women, men, and children toiled and produced rice crops on nearly seven hundred acres on Gowrie Plantation, often with the heavy pressure of the overseers and managers. Louis Manigault, the owner, boasted that "the truth is on a plantation, to attend to things properly requires both master and overseer." In the thickly cultivated fields of the South, the overseers served as an arbiter of mutinous toil. They would sweep through the fields, observing, commenting, and in some cases whipping the enslaved who failed to work at a quick pace. The fear of retribution (or being the direct target of physical assault) was part of how the plantation sustained subjugation.

Throughout his life, Manigault never worried about money, given how much land and how many people he owned. Louis Manigault inherited three plantations: Silk Hope, Gowrie, and East Hermitage. Of these, the Gowrie Plantation was the most grisly. Between 1847 and 1854, about 90 percent of the children on the Gowrie Plantation died before they reached the age of sixteen—a number that does not account for stillbirths or miscarriage. In some cases, their deaths were worth more than their lives. During this period, some enslavers insured their slaves in case of unforeseen injury or death. The Manigault family could receive as much as $44,000—the value in 2022—if one of their slaves died. The commercial examination of Black life casts a chill, a sense that some people could have more value in death than they did alive.

In the fall of 1854, dozens of the enslaved people on the Gowrie Plantation suddenly grew tired. Some of them expressed abdominal discomforts, others suffered from incessant diarrhea. Unable to work, many of them were bedridden from cholera. Louis Manigault wrote to his father about the people who died at the Gowrie Plantation:

Here they are in the order in which they died. —Hester, Flora, Cain, George, Sam, Eve, Cuffy, Will, Amos, Ellen, Rebecca, —Eleven from Cholera, and two Children viz.: Francis and Jane not from Cholera. —In all Thirteen names no longer on the Plantation Books.

These names document a tragedy, but for Manigault, the plantation manual was a till where he listed his sentient property. These deaths were not an aberration. Between 1852 and 1859, at the height of the third cholera pandemic, hundreds of thousands of people in Asia, Europe, North America, and Africa died from cholera. Most of these transmissions were rampant in cities such as New York City or London, in the working-class districts of industrial centers. But for the enslaved at the Gowrie Plantation, cholera was another scourge in the Georgia landscape.

When another cholera outbreak occurred in May 1856, Manigault was even more agitated. "Considering the immense losses we have experienced during the past three years, the Cholera having off in 1852 and 1854 many of our very best hands, a destructive freshet visiting us in August 1852, just in the midst of harvest." In this moment, what mattered most for Manigault was maintaining a profit, which was something he was easily able to sustain. In 1861, at the beginning of the US Civil War, the Gowrie Plantation was worth $266,000, a value that reveals that even during a political conflict he faced little adversity.

Manigault's manual served many purposes, but one of its main concerns was to provide an apt, but unsettling, portrait of the cost of being a slave. His guidebook also reveals the level of organization required to keep a person captive from the perspective of an enslaver. Compiled between the 1850s and 1870s, his notes include information on plantation life, rice cultivation, market conditions, and enslaved people (and later sharecroppers after the Civil War). The manual expressed all the ways Black captives' lives were cut short not just through arduous labor but contagion on the plantation. For many nonliterate enslaved people, proof of life appeared in plantation manuals, sale records, and broadside ads asking for their reward. They lived in congested quarters and worked until exhaustion, so the bodily damage probably felt like hell on earth when they were infected with cholera.

For those infirmed with cholera, the experience of the illness looked like this: It took a grip on their body, becoming more prominent than life but at once also stirring one's strength. Slightly rumpled, a cholera sufferer might suffer from lethargy and dehydration, given how briskly liquid and solids would be expelled from their body. They might endure vomiting, muscle cramps, sunken eyes, wrinkled skin, ruptured capillaries, and, if left untreated, death. The disease's essence extended far beyond the body. As the author Susan Sontag tells us, "Cholera is the kind of fatality that, in retrospect, has simplified a complex self, reducing it to [a] sick environment." For the enslaved people of the Gowrie Plantation, the densely packed cabins were a recipe for mortality.

Water played a significant role in how cholera was managed on the plantation, which was further evidence of how confinement—in an unsanitary facility—was a breeding ground for cholera. Contaminated water—which could be vaulted in a well—was a medium that fueled cholera's carnage. Yet, cholera can also be contracted by ingesting food sullied by excrement that harbors the bacteria. The lack of proper sanitation for Black captives on the plantation occurred because their enslavers didn't think they were worthy of adequate disease management.

The deaths from cholera were not exceptional and even by admission of the enslaver, were due to the unhygienic housing conditions.

But these deaths were amplified by the abominable mold that multiplied on the windowsills of the slave cabin. With Louis Manigault noting, "Everything is nasty & dirty about the [slave] settlement . . . we have no more time now for this year to white wash & all now remain dirty & dingy until next year . . . everything is covered with the freshest sediment & the fields 'Stink'!" Although he recognized and had the power to address the problem, he did little to improve the structural environment, which could have limited the spread of cholera and other infectious diseases.

At times, Manigault remarked—with a patronizing tone—about his alleged fondness for some of the enslaved men on his plantation. Though he rarely uttered positive remarks about the enslaved, he expressed sanguine comments about the ones he had direct contact with, including Jack, another enslaved person who was his "old play mate." In the American South, enslavers sometimes mentioned their enslaved as an extension of their family, even though Black people were rarely given the same rights as white Americans. Manigault's papers leave no impression of whom the enslaved loved or what moved them. For Manigault some of the Black men were attractive and amicable; they were made available to him. One enslaved man, Stafford, was described by Manigault as "the finest looking Negro I ever saw." But I was curious: Did Stafford and Jack have indulgent memories of him? Manigault's fondness for Jack and Stafford was limited, given that he believed that "the only suitable occupation for the Negro is to be a Laborer of the Earth, and to work as a field Hand upon a well-disciplined plantation." This does not mean that they were respected, per se, rather, that the enslaved person was only useful so long as they "behaved" according to the enslaver's will. They tried to find any reason to justify this way of life, even if it meant concocting "remedies" that would provide basic therapy, but not question the institution that made their captives sick in the first place.

The professed fondness that Manigault expressed about Stafford materialized into his attempt to treat the man for cholera. In 1852, Manigault set aside an herbal regimen for Stafford. The mixture included: twenty grams of calomel and two grams of opium into a powder. According to his records, the dosage was given every two hours, followed by castor oil six hours after the last dose. The fact that the prescription was specifically assigned to Stafford suggests that the enslaver wanted to alleviate his symptoms, possibly to minimize the loss of his property, but most likely to ensure that there was an additional stalwart hand to pick the harvest. But given the physiological portent of the two compounds—calomel and opium—the reagent was unlikely to work. The former is a metabolic poison that, in high doses, can kill a person. Today it is used as an insecticide. The latter is commonly prescribed to relieve pain, though it is highly addictive. It is unclear whether Stafford consented to the medication or if the reagent provided temporary relief. What is certain is that Stafford died soon after the alleged treatment.

On the surface, it looked as if Manigault "cared" for Stafford and other enslaved people who were provided cholera treatment. But given the ingredients, it most likely triggered side effects such as vomiting or even death. What cannot be lost is that Stafford was living in a confined place, a section of the plantation in Manigault's property. What Manigault's manual reveals is an inexpressible silence, the gaping hole in Stafford's narrative. He doesn't furnish any lurid details about Stafford's childhood memories, decaying body, or how he drifted closer to expiration. The recognition of Stafford's life, illness, and eventual demise is mediated by the person who made him captive. Encasement was not the only pre-existing condition that led to death from this epidemic. Cholera's wrath was global.

During the nineteenth century, the battle to fight cholera was asymmetrical. Cholera marked life and death in mid-nineteenth-century America; for example, there were 40,000 reported cholera deaths in June 1850 alone. Well into the middle of the third wave of the cholera pandemic in the nineteenth century (1846–1860), two US presidents perished from the disease. President James K. Polk died of cholera morbus in 1849, and his successor, Zachary Taylor, died one year later. Although the presidential deaths revealed that even elite people could die from the disease, the power to confine oneself and have access to proper sanitation—unlike on a plantation—improved one's chances of avoiding and surviving the outbreak.

In the American South, cholera struck with unsparing precision, though how death was recorded was dependent on one's status and access to freedom. We know about the death of some enslaved people—such as those living in forced captivity at Gowrie Plantation—because their enslavers saw a financial reason to account for their death. (Given that it was illegal to instruct a slave on how to read or write, most could not document their lives through text.) While elite and free persons had their death recorded in state mortality records, enslaved people were listed in the census by their enslaver, not as their individual selves.

Many scientists seek to mark their place in history. Some do so out of hubris, others do it to save sections of humanity. By the middle of the nineteenth century, scientists were on a race to find the cure and adequate treatment for cholera: Their approach was to isolate the infected and observe the biological entities that made them different. Like many ambitious anatomists, Filippo Pacini tried to find the cause of cholera. Trained in medicine in Italy, he spent the early part of his career studying the human nervous system. While he was a professor at the university of Florence, the cholera epidemic struck in 1854. Pacini belonged to a minority of physicians who had access to living cholera victims and cholera-ridden corpses, so he redirected his attention to investigating the infectious disease. After collecting the fecal matter of several patients, he concluded that a microbial agent, of the genus *Vibrio*, caused cholera. Although his research went virtually undetected by non-Italian scientists for several decades, other scientists provided sweeping public health interventions to mitigate cholera's transmission. Outside of the plantation, emerging scientific theories impacted the ways scientists responded to disease management in cities. A phenomenon which was scant in the countryside.

Like Pacini, the British scientist John Snow worked to find the source of cholera, but rather than do so in a laboratory, he mostly focused on the gritty streets of London. On a mild day in late August of 1854 in London, John Snow started counting the dead in Soho. Walking through the semi-deserted alleys near Broad Street, he learned that the brewery workers were some of the few who survived the recent cholera outbreak. While most people thought the people who died from cholera did so because of bad air or soil, he rejected those theories and wondered about other possibilities. Unfazed, Snow focused on the living rather than the dead. In another section of town, he found that the brewery employees at the Workhouse on Broad Street received their water from Grand Junction Waterworks, while most cholera victims received their water from the (now infamous) water pump. As his investigation expanded, Snow found that this was among other water sources that were linked to cholera transmission.

Although Snow never identified the microbe responsible for cholera, when he established that contaminated water was the mode of circulation, he had enough evidence that dispelled the theory that cholera spread through the air. Today, the Broad Street Pump sits on Broadwick Street in London as an homage to John Snow and his ingenuity—a catalyst for public health officials in London to act and sanitize the water sources in the city. The knowledge and motivation to improve the city's water quality was seen as a public good and eventually reduced the spread of cholera and other infectious diseases throughout the city. For enslaved people living on the plantation, the initiative to rid their surroundings of infection, as was the case at Gowrie Plantation, was sorely missing.

Cholera had a different denouement in plantations than in free society. On the plantation, the disease could sweep into the freezing cold sleeping quarters or in the long days of the sultry summer. Enslavers sought advice from doctors during plantation outbreaks, or in some cases, they consulted medical pamphlets that specifically focused on so-called "plantation medicine." What they considered sound medical advice was anything but.

There are many ways to look at the plantation. One practice is to see it as a commercial enterprise based on the forced labor of one group of people to accumulate the wealth of another group. Another way to view it is as a torture camp predicated on what Professor Saidiya Hartman asserts "was an assemblage of extreme domination, disciplinary power, biopower, and the sovereign right to make die." So long as these people were enslaved, how could they avoid getting sick in a broken world? Louis Manigault's recipe for Stafford was proof that they tried to assuage the disease through a reagent, but his actions show that he was invested in the false promise of plantation medicine in Georgia and other southern plantations.

The False Promise

Well before European colonists arrived at the Mississippi River, the Choctaw people were one of the major ethnic groups that considered the area their home. For the first decade of the nineteenth century, the

Choctaw hunted deer and traded livestock with European settlers. By the 1830s, the US government introduced the Indian Removal Act, which meant that Choctaw and many other indigenous groups had three years to relinquish their ancestral land and move west of the Mississippi River. There are countless names for this forced migration, but when I learned this history, I was told that this was the Trail of Tears. The continued expulsion of Native Americans from their communal land was a bulwark against their sovereignty.

Well into the 1850s, on the patches of southern land on both sides of the Mississippi River, there were port cities that shipped cotton, goods, and Black captives to larger cities or plantations. Beyond the marsh ruins, in plain sight, the neighboring land in Memphis was converted into plantations by white American proprietors. One of those men was John Pope, a southerner who settled in West Tennessee by way of Alabama. After graduating from Yale University, Pope served as a legislator in Alabama and president of the Shelby County Agricultural Association. Urban-based enslavers like John Pope did not have to oversee the daily activities on their estates. Pope relied on the assistance of punitive men to chastise the enslaved. With over seventy enslaved people working on his farm, his cotton plantations were valued at over $100,000 at the time. Pope's concern wasn't just relegated to his profit margins, he was also attentive to the anatomy of the plantation.

Plantation medicine, as practiced by men like John Pope, was an uncomfortable space where he maintained he could heal those who were held captive, yet there was little evidence that his methods were effective. In some cases, he spent money for a trained professional to heal a slave, but mostly he took matters into his own hands. By the time he settled near Memphis in 1848, he believed that he had found the cure for cholera.

Given that the 1833 cholera epidemic had ravaged most of the

Eastern coast of the United States, he decided to concoct a "remedy" which included "a spoonful of spirits of camphor with two of essence of peppermint and 25 drops of Laudanum" to the slaves who contracted cholera on his plantation. In essence, the laudanum, a mixture of opium and spirits, helped to relieve pain, but it wasn't necessarily adequate in obliterating an infection. In his hubris, Mr. Pope claimed that he "never lost a patient by using this remedy." So many of enslavers' opinions about cholera were rooted in their hubris about contagion, "commonsense" observations that meant that anyone with some organic or not-so-organic reagent could claim to be a healer. He was not alone.

In 1819, one of the best-selling books in the United States was a plantation manual, Dr. James Ewell's *The Planter's and Mariner's Medical Companion*, one of the popular medical texts of the time. Published in 1807 at the price of $50—equivalent to $1,500 today—and dedicated to Thomas Jefferson, the manual argued that cholera was caused by "redundancy and acrimony of the bile—indigestible food or such as become rancid or acid on the stomach—poisons—strong acrid purges or vomits—passions of the mind, or a sudden crack of perspiration." According to this theory, the cause was both a mixture of common cholera symptoms—vomiting and perspiration—and spurious "passions of the mind." His understanding of the disease's origin was off the mark and his true intention was to establish a concoction. In reality, his remedies were based on his (ill-informed) perception of the malady. Dr. Ewell recommended consuming a large volume of chicken water, taking a warm bath, and ingesting a small quantity of nitrate or castor oil. The regiment was intended to expel the disease from the body, but if that did not work, the enslaver might resort to bleeding the enslaved person—an extension of a medical practice.

To some credit, even when manuals were shrouded with (false) biological essentialist ideas about the enslaved, there was recognition that the labor they were subjected to caused them great harm: "That the constitution of the African is more firm than ours, and better fitted to sustain the toils of warm climates, is very certain, but it is equally true that his daily labors, with the sudden changes of weather, often put his constitution, good as it may be, to trials, which loudly call for every aid that humanity can possibly afford him." On the one hand, these men held on to firm views that alleged biological differences between races, but on the other hand, they were conscious that Black captives experienced major distress from enslavement. Acknowledging this setback, some enslavers found other strategies to address health inequities.

When it wasn't assured that the instructions from the plantation manuals would work, enslavers did occasionally employ white physicians or Black midwives on an annual salary basis to treat the enslaved. But the ability to provide aid, both in name and practice, was an unbearable space that revealed the paradoxes of providing care under confinement. The cost of a physician could be as high as $200 for a visit, which could include treating the enslaver and the enslaved. But in most cases, enslavers did not have access to doctors, nor did they necessarily seek professional care to help treat a sick enslaved person; more often, they relied on home remedies to treat the ill.

Precisely because of the taxing living conditions on a plantation, medicine had its limits, but Black healers tried their best to work under these circumstances. For the enslaved and the manumitted, medical care was far more receptive when it came from another Black person. In his memoir, *Narrative of the Life of Frederick Douglass, an American Slave*, the African American abolitionist Frederick Douglass was discerning, recognizing the mental dissonance shaping his enslavement: "Born for another's benefit, as the *firstling* of the cabin flock I was soon to be selected as a meet offering to the fearful and inexorable demigod, whose huge image on so many occasions haunted my childhood's imagination." But at the same time his words made clear that the plantation, even if it was a site of torture, contained healers drawn from the ranks of the enslaved themselves. People like Uncle Isaac Cooper, who later added the prefix "doctor" to his name, were among the enslaved persons who provided medical care to other slaves. Speaking of Uncle [Doctor] Isaac Cooper, Douglass remarks: "He was always on the alert, looking up the sick, and all such as were supposed to need his counsel. His remedial prescriptions embraced four articles. For the diseases of the body, *Epsom salts* and *castor oil*; for those of the soul, *the Lord's Prayer,* and *hickory switches!*" He listened intensely to their distress and pacified their ailments, without condescension.

Part of the reason that these medicinal practices were homegrown is that the plantations were almost uniformly located in regions that lacked public health facilities. While medical pamphlets were supposed to be their quick fix, the chattel slave economy failed to create robust medical institutions for the majority of its population.

If living was saturated with torture and poor health, some enslaved people found a way to escape. They were burned by the vicissitudes of the plantation, the challenges of cholera, and for Black women, the burden of pregnancy and gestation against their will. Much of the abuse that Black women faced was tied to sexual harassment and assault, which meant that any form of refuge was better than slavery.

Harriet Jacobs

When Harriet Jacobs was born in 1813, North Carolina had a one-hundred-year-old slave code which declared that whenever an enslaved person left a plantation, they had to carry a document from their enslaver stating their purpose and destination for travel. This same code forbad enslaved people from gathering in groups for fear that their communion would spark a riot. Born near Edenton, North Carolina, on a plantation in the Albemarle Sound, she was the daughter of Delilah Horniblow and Elijah Knox, two enslaved people—which meant she was born unfree. What set her apart from most enslaved people is that she was eventually able to register her life through her own words and text. Margaret Horniblow, her mother's owner, taught Harriet how to read and write—even though the North Carolina Assembly prohibited anyone from instructing enslaved people to be literate. Southern plantation owners feared that educated Black people could threaten chattel slavery, by empowering themselves and inciting more insurrections. Nevertheless, Jacobs punctures through the sinister aspects of captivity by laying bare the onerous trials common for enslaved women during the mid-nineteenth century.

By the time Harriet Jacobs published *Incidents in the Life of a Slave Girl*, she did so under a pseudonym—Linda Brent. Written between 1853 and 1858, the text is swarming with sordid details, casting insight on the perennial humiliation and mortal consequences of sexual assault Black women faced by their white male enslavers. Worn down by the jolt of unending desolation, Harriet described Black maternal mortality: "I once saw a young slave girl dying soon after the birth of a child nearly white. In her agony she cried out, 'O Lord, come and take me!'" Her pleas shrilled through the air, drubbing the lesson to everyone else that Black women could have their life taken away simply for birthing a child.

During her lifetime, Harriet occupied many lives during enslavement and manumission: She was a daughter, mother, abolitionist, and author. The loving relationships she had were predicated on recognizing the barbarism of chattel slavery. When her father passed away in 1826, she blamed his death on his enslavement noting, "My father, who had an intensely acute feeling of the wrongs of slavery, sank into a state of mental dejection, which combined with bodily illness, occasioned his death when I was eleven years of age." Soon after, her mother died, a similar grief occurred. However, by this point, she was transferred to another plantation. For as long as she could remember, her labor, movement, and body were all monitored because her life wasn't hers.

Most of us have a breaking point, a juncture where we can no longer bear the circumstances that have become habitual. But Harriet—by her own account—revealed that instance when Dr. Norcom harassed her. One of these points came when, at the age of thirteen, Dr. James Norcom, her de facto enslaver, tried to sexually assault her. He was forty years her senior. Jacobs notes, "He peopled my young mind with unclean images such as only a vile monster could think of. I turned from him with disgust and hatred. But he was my master." His harassment did not stop with his sexual advances; they included physical and emotional distress, such as barring her from seeing her children or touching her without consent. But the cruelty wasn't just in the act, it was knowing that at any moment, one could be terrorized without any form of retribution.

In the late 1820s, shortly after Dr. Norcom denied Harriet her request to marry someone she loved, Jacobs wrote: "He sprang upon me like a tiger, and gave me a stunning blow." Incensed and dumbstruck, she took a moment and pulled herself back and resisted his encroachment. Rather than let him be absolved of repudiation for his assault, she chided: "You have struck me for answering you honestly. How I despise you." What unfolded was a staggering rhythm of slurs: He threatened her with death. In response, Harriet proclaimed that she wished she were dead; then, he warned her of imprisonment, and she rejoined that jail would be better than the plantation. Harriet expressed a raw confession, but it is also a testament that she would not let her tormentors have the final word. Like many enslavers, he cultivated a hostile world, and no matter how much Harriet protested, he was insistent.

Harriet's torment was not merely due to her physical state but what it meant for her to be bonded and pregnant at the age of fourteen by Samuel Tredwell Sawyer, a young lawyer in his early thirties. For weeks, she became incapacitated. She lay in bed "too ill in mind and body." The infirmity was not from contagion but from the prenatal symptoms that arose—fatigue, nausea, bloating, and constipation. Most pregnant people experience these conditions, but for an enslaved person, they are especially gruesome. Like many enslaved women who knew their children could be taken away from them, she reckoned that death was better than slavery. After giving birth, what gnawed at her soul was knowing that the child was not legally hers, and like her, would be born into slavery. She "often prayed for [her own] death." Harriet didn't just want him to die; she wanted them both to die. For a while, Harriet thought this would be possible, given that her son, Joseph, was four pounds when he was born, and both of them spent several weeks afterward with "chills and fever." Even the physician, Dr. James Norcom, her enslaver and tormentor, didn't think he would survive.

As a plantation owner, Dr. Norcom behaved in a way hewed to the tenets of slavery—transgression, assault, and abuse. He possessed the land, the resources, and the power to heal the people he had enslaved, but in reality, his legal jurisdiction over them meant that he had control over everything—including their children. Dr. Norcom assisted in the delivery of Jacobs's child, but the care she and her son received postpartum came from her grandmother and the other Black people in the surrounding plantations. Part of the contradiction of his role as a physician who enslaved others is that his authority to cause harm eclipsed his duty to heal. This was not lost on Harriet. She realized something even more complicated. Some healers, who were meant to provide care, were not only complicit within the plantation system but were also meant to ensure that the sentient "property" could only live under one condition: dejection.

The plantation, and chattel slavery as a whole, created a paradoxical situation regarding the enslaved women's form and constitution. The dangers of pregnancy were so tenuous as to make their survival and that of their fetus less feasible than for free white women. They were robbed of the comforts of rest, sustenance, and protection. Being held captive, there was no legal ground for them to access proper medical attention. There was little to no relief from daily labor, so when she icily exclaimed that her death and her infant son's would be welcome, she was also aware that this would mean manumission from forced labor.

African American scholars and writers of slavery and its afterlives, such as Saidiya Hartman, have deepened how we think about these histories, drawing on the perils of plantation life and the fortitude that enslaved women had even in reproductive bondage. Published in 1997, Hartman's *Scenes of Subjection* argued that Black freedom was unrelenting because racism was so ruthless. Hartman's research—in particular—shifted the US American pedagogical terrain. By reckoning with the moral gravity of enslavement, that is, she read with and against the archive to find "the relationship between subjectivity and injury" within the context of chattel slavery. That injury boiled down to sexual assault, forced pregnancy, and maternal mortality. Time and again, enslaved women such as Harriet Jacobs exercised patience and adaptability, despite enduring sexual assault.

Even when Black women were the target of rape, they were steadfast in exercising their agency. One area in which a Black woman had autonomy was intervening in gestation, terminating a pregnancy, or enacting infanticide. As Angela Davis notes in *Women, Race, and Class*, "Abortions and infanticides were acts of desperation, motivated not by the biological birth process but by the oppressive conditions of slavery. Most of these women, no doubt, would have expressed their deepest resentment had someone hailed their abortions as a stepping stone toward freedom." When Black women induced abortion and infanticide, they exercised control where there had been none. But more importantly, they also interfered with their enslaver's profiting from their reproductive labor.

As Historian of Medicine Deirdre Cooper Owens remarked, successful births within the slave community were felt as triumphs for some and a tragedy for others. For the enslaver, they would have another worker. For an enslaved person, the feeling was mixed; some people valued their kin and wanted their children to survive, but some, like Harriet Jacobs, were terminally conflicted, knowing they would be denied liberty. And enslaved women's fertility played out in material ways. Those proven to bear offspring were worth as much as 25 percent more than those who could not gestate.

Although Jacobs's account of her infant's illness and postpartum survival is brief, her captivity speaks volumes about the morbidity that arises in confinement; her memoir details her journey into the plantation, how the environment was abrasive, and how the enslaved did their best to aid each other to full recovery. Jacobs's text bears affinity to the other freed writers who narrated confinement, sickness, and death on the plantation. The testimony of the enslaved shows how they endured the violence of their confinement on the one side and illness on the other, culminating in the idea of the plantation as an incubator of resistance. To Jacobs, abolition was the only answer to a brutal system, while the other solution was death, noting, "At least death frees the slave from his chains."

In this social context, where Dr. Norcom, a physician by trade, Harriet's de jure owner since she was a child, was tasked to be a healer, he was duty-bound to help her with her delivery, but given that Jacobs's children's father was Samuel Sawyer, a white lawyer who was not her enslaver, Dr. Norcom expressed disdain about her affair. Initially expressing guilt for having sex with Mr. Sawyer, Harriet later pointed out: "Still, in looking back, calmly, on the events of my life, I feel that the slave woman ought not to be judged by the same standard as others." Her pregnancy was unexpected but not uncommon, given that many Black women had little choice about whether or how they would have children. Nevertheless, the success of Harriet's pregnancy mattered because Black women like her were responsible for reproducing the next generation of slaves. Enslaved women who became pregnant faced little medical care on the plantation, often perplexed by a system that left them with little choice but to dedicate themselves to a fruitful bonding process with their offspring.

Pregnancy was one of the most life-threatening conditions possible during the nineteenth century. While estimates of maternal mortality varied based on class, race, and region, some estimate that less than 1 percent of births resulted in maternal mortality. Being a pregnant enslaved woman was even more dangerous—often accompanied by premature births, underweight infants, and ongoing fatigue for the mother and child. At four pounds, Joseph's weight was not a deviation but indicative of the harmful effects of Harriet's bondage. As Dorothy Roberts observed in *Killing the Black Body*, "Healthy pregnancy was hardly possible with the strenuous labor, poor nutrition, and cruel punishment bonded women endured."

Roberts notes that Black enslaved women were more likely to experience infertility and miscarriages, a mirror to the reproductive lives of African American women today. This reproductive subjection, which often came with pregnancy loss or maternal mortality, on a plantation was not an anomaly but the norm.

But the thrust of what Harriet describes extends far beyond reproduction. As she wryly notes, the plantation not only created the conditions that fueled psychological, physical, and sexual assault but also incited a spark for her resistance. Jacobs weaves her desire to escape throughout the text, stating "until I came into the hands of Dr. Flint [pseudonym for Dr. James Norcom], I had never wished for freedom till then." We often think of freedom in simple terms, as the state of not being captive in general, and yet, it can also surface in other ways. Freedom can be a subtle refusal, focused on one pariah that irks the soul.

Gender-based violence and sexual assault were common, but it was minted into the fabric of the plantation, often used to coerce enslaved people to bend to a master's will. Dr. Norcom's leverage was substantial because he had power over Harriet and her child. After years of agony, Jacobs was tired of his predation. But given his assertion, "You are my slave and shall always be my slave. I will never sell you, that you may depend on," *she had to leave.* She no longer wanted the scrutiny of Dr. Norcom's gaze or his legal claim over her life and body. By 1835, Harriet escaped to an ad-hoc haven near the person she loved the most—her grandmother. So, Harriet anchored herself in her grandmother's cabin and hid in the modest crevice of her attic. Molly Horniblow, her grandma, was emancipated so she was able to have some control over her home. Being nearby, and remaining in this liminal state, meant that if the lives of her two children, Joseph and Louisa Matilda, were threatened, she could quickly resurface if necessary and at least protect them. Harriet noted,

I suffered for air even more than light. But I was not comfortless. I heard the voices of my children. There was joy and there was sadness in the sound. It made my tears flow.

The attic was her compromise, an act of control where she would be within hearing range of her children.

To be held captive and subjected to sexual advances by a man whom you detest is to live in anguish. But rather than be a slave, Harriet became a phantom who moved from one state of captivity to another. She had no large furniture, nor could anyone else fit into these quarters. At nine-by-seven-feet by her measure—the garret was better than Dr. Norcom's sexual advances or being subject to daily acts of punishment. For seven years, she confined herself to this attic, which she called "the loophole of retreat." She enclosed herself into a space where she could breathe, write, and sleep.

Slavery was brutal as much as it was irrational; it posed an existential threat to Black families as much as it did to Black women's bodies. Although she escaped from her predator, there was always a possibility that her concealment could be breached. But the risk was worth it. Between 1810 and 1850, approximately 100,000 people escaped from US chattel slavery, and in 1842, seven years after burrowing into her grandmother's garret, Harriet left the American South for good. The scholar Katherine McKittrick writes about this conundrum: "geographic options such as escape, concealment, and racial-sexual safety can be, as Brent [Harriet Jacobs] explains, bound up with troubling spatial strategies." Concealment was dreadful, but anything was better than prolonged misery. Up North, she stood better chances of having autonomy over her body.

When I first read Jacobs's testimony, I was flooded with a range of emotions, including the heavy weight of knowing that her enslaver, by design, did not have her physical well-being in mind.

Regarding the moral code of physicians, Hippocrates wrote in his treatise *Of the Epidemics*, "The physician must be able to tell the antecedents, know the present, and foretell the future—must mediate these things, and have two special objects in view concerning disease, namely, to do good or to do no harm." Although the exact phrase of "do no harm" has been contested by scholars since antiquity, the message was clear for most physicians in the West, that the primary purpose of physicians was to avoid, as much as possible, causing injury intentionally. Dr. Norcom was able, somehow, to overlook the wound he inflicted on Harriet, extracting labor without compensation. When he sold Harriet's children and profited from the sale, he tormented her. When he physically and sexually assaulted the enslaved people on his farm, he abused them. By most accounts, Dr. Norcom was a pariah to Harriet Jacobs. The act of confining a person, and the brutality and vulnerability they faced, meant that in mind, body, and spirit, suffering was not only present, but it was ceaseless.

When Harriet initially arrived in the North, she worked in New York City, as a nanny for Nathaniel Parker Willis, a white poet and editor. Eventually, she became acquainted with other abolitionists, some of them encouraging her to practice her literary skills. Freedom saved her; reading nurtured her. But what mattered most was that her confidence thickened: "The more my mind had become enlightened, the more difficult it was for me to consider myself an article of property." That self-assurance was stunted by a compromise. When northern and southern states could not agree on whether to abolish slavery, Congress settled for the passage of the Fugitive Slave Act of 1850, which required escaped slaves to be returned to their enslavers even if they lived in a state where slavery was abolished. The retribution: People assisting escaped slaves could face a fine, and the self-manumitted might face whippings, branding, or amputation after they "returned." Like many newly liberated people in the North, Jacobs shuddered at the thought that she could be fetched, chained, and returned to the torture camp.

By the end of Harriet's life, slavery in the United States was abolished. Freedom gave her the space to organize, living as she wanted, but it also settled her mind and placed less stress on her body. She was living in the Hudson Valley in upstate New York, far from the cramped quarters of her grandmother's attic.

For many African Americans, enslavement left many freed people neglected, tired, and broken—and the physical susceptibility to ill health was not just about reproduction, it was the fact that anti-Black discrimination continued to undermine the sovereignty of people who were enslaved, even well after they were free.

Live Free or Die

After emancipation from slavery, most African Americans carried a massive objective—to make freedom come into full fruition. W. E. B. Du Bois captured this beautifully when he noted: "The very feeling of inferiority which slavery forced upon them fathered an intense desire to rise out of their condition by means of an education." When the US Congress passed the Freedmen's Bureau Bill on March 3, 1865, the newly formed government had a heavy task: to ameliorate the physical and social well-being of the four million emancipated people. The US Civil War had just ended, and for the first time in American history, African Americans were elected to local and federal government. Many enacted social programs intended to subvert the aftershocks of the nation's divisions. The striking and revolutionary program to provide free, universal education was one of the principal programs that emerged during this period. Another core tenet was to provide free healthcare to Black Americans.

For many African Americans, the ghastly side of poverty had a chilling effect on the body, so it made sense to invest in the health of the emancipated. One congressional report stated it was essential to "extend medical aid as far as practicable to refugees and freedmen who became sick and were unprovided for by any local supply from a private source." Within a few months, they created a government body that brought this directive into realization. Founded in 1865 by African Americans, the Freedmen's Bureau provided relief for formerly enslaved people. Meanwhile, the Medical Division cared for Black Americans. Even if they were understaffed, with only 120 physicians to care for the entire South, they believed everyone was entitled to free medical treatment.

Similar to education, their goal during Reconstruction was establishing healthcare for all. They documented and treated infectious diseases such as cholera and smallpox for seven years. Although their ability to carry out services varied depending on their location, they tried their best. Many African American medical workers petitioned the federal government for structural assistance, recognizing that medical care required professionals in every region of southern society. Rev. A. S. Fiske, a member of the Freedmen's Bureau in Mississippi, lamented: "The proper care of the sick and charge of the sanitary affairs requires that each provost-Marshall district should have a medical officer who should be in control of all sanitary affairs on the plantation." These initiatives were imperative because they revealed that health was not solely characterized by individual treatment but by a public health initiative.

Soon after the US Civil War, exasperated and hardened, formerly enslaved *African Americans* flocked to the capital city, hoping to fulfill the promise of freedom. Washington, D.C. became a haven for a new wave of refugees, African Americans who fled the southern plantations, hoping to find work, education, and dignity. They reconstructed themselves into the vicissitudes of the city, scrambling to learn and love as the newly liberated. Several years later, universities were tasked with training a new layer of Americans in a multiracial setting. In a terrain that had harsh boundaries of humanity. But slowly, newly created institutions had to reckon with a new era of equality.

When the semester began on November 5, 1868, Dr. Lafayette Loomis addressed eight medical students at Howard University.

What a field of honorable toil is here! How limitless its opportunities for good! How worthy the life that uses them well! Such toil, opportunities, and honor open to the patient, conscientious, and faithful student of medicine. May the after years of your lives my young friends, justify the hopes of the present hour, and along your sometimes weary student's life may you never forget that success comes only of patient toil, and that patient toil never fails of success.

Seven of those students were African American, and one was a white American. Howard University was initially founded to train formerly enslaved people in theology but later expanded to educate people of all races and genders across various fields, including agriculture, law, and medicine. Entry was predicated on the requirements of the time: evidence of superior moral character, adequate English language comprehension, elementary mathematics skills, and proficient knowledge of Latin vocabulary.

In 1870, medical students were instructed in the classroom and out of the newly created Freedmen's Hospital, a product of Reconstruction's success in providing free medical care to Black Americans. Founded in 1865, the Bureau of Refugees, Freedmen, and Abandoned Lands, or the Freedmen's Bureau, was tasked to redistribute relief and rations to formerly enslaved people and their descendants. The Bureau for Freedmen—one of the many components of the Reconstruction era—wasn't just about protecting Black people's rights and establishing a transition to Black equality in various professions. African Americans established the Freedmen's Bureau for the Ministry of Health, which improved the lives of a generation of Black Americans and encouraged them to get medical care and medical education.

Several years after its founding, in 1871, five medical students graduated, two African American—James L. Bowen and George W. Brooks—eventually providing services to Black Americans. By 1877, Eunice P. Shadd became the first African American woman to graduate from Howard University's College of Medicine. (Mary D. Spackman, a white American, was the first woman to graduate from the institution in 1872. Rebecca Crumpler was the first African American woman to earn a medical degree in 1864—though she earned her degree at the New England Female Medical College, now Boston University.) The accumulation of students is noteworthy because it normalized the idea that Black men and women could be doctors, a concept that was unfathomable decades before.

Between 1871 and 1896, Howard University alone awarded over 450 people a medical degree—most of whom were African American (men). But they were not alone. Altogether, Black medical schools in North Carolina, Louisiana, and Tennessee—most of which emerged during the Reconstruction era—trained 1,465 people between 1870 and 1910. In many cases, these individuals were the first in their families to attend university, let alone have access to a professional class. Some of them were born free, and others were born enslaved. Their education, on the whole, was made possible by a part of American history's period to heal its historical wounds after the Civil War and establish steadfast institutions that would provide care to the newly incorporated citizens of the United States, African Americans.

Hailing from Boston's neighborhood of Beacon Hill, Dr. Rebecca Crumpler, an African American physician who was employed by the Bureau, wrote about her experience of working with formerly enslaved people in Virginia recalling the large volumes of patients needing care: "I was enabled, through the agency of the Bureau under Gen. Brown, to have access each day to a very large number of indigent, and others of different classes, in a population of over 30,000 colored [people]." Some of the Black-led departments had to deal with underfunding and recalcitrant whites, who fought to undermine the rights of Black people to access free healthcare. The ambition to expand healthcare access was thwarted by money and anti-Black racism.

The gains made in the latter part of the nineteenth century began to deteriorate at the beginning of the twentieth century. In 1906, Du Bois's "The Health and Physique of the Negro American" argued that health disparities between people of the Black and white races were caused by the unfulfilled promises of Emancipation to provide relief or restitution to Black Americans rather than biology. Freedom offered the formerly enslaved the possibility to learn, grow, and heal, but without land, resources, or safety, it was nearly impossible for them and their descendants to flourish with vitality.

In 1910, that started to change when Abraham Flexner, an American educator, published "Medical Education in the United States and Canada," a journal article that evaluated medical schools in North America based on admissions, laboratories, and clinical material. On the surface, the reports could have been a catalyst to provide funding and resources to institutions that were falling behind, especially those that lacked access to hospital teaching facilities. But in reality, his report was the death knell for two-thirds of American medical schools at the time. After visiting institutions in Europe and the United States, Flexner noted, "We may safely conclude that our methods of carrying on medical education have resulted in enormous over-production at a low level, and that, whatever the justification in the past, the present situation in town and country alike can be more effectively met by a reduced output of well trained men than by further inflation with an inferior product." At the heart of his article was a desire to establish a standard for medical education that would restrict the number of physicians trained in the United States. The Flexner Report was a turning point in the progress achieved for Black health workers in the nineteenth century. Five of the seven historically Black medical colleges were closed down within a decade of its publication. At the time, those remaining were Howard University Medical College and Meharry Medical College, a predicament that could still impact how many Black physicians are trained in the United States.

For African Americans, the failure of the government to provide retribution has been a long battle, incommensurate with the narrative that legal emancipation was enough. In 1964, a century after Black people were emancipated from slavery in the United States, the Civil Rights leader Malcolm X interjected and quickly rebuked the idea that the United States had overcome its issues with racial inequality. He noted, "If you stick a knife in my back nine inches and pull it out six inches, there's no progress. If you pull it all the way out, that's not progress. The progress is healing the wound that the blow made. And they haven't even begun to pull the knife out, much less try to heal the wound. They won't even admit the knife is there." What Malcolm X affirmed is that simply ending the problem is insufficient—it must be reorganized.

No matter how much African Americans worked to undo centuries of Black captivity, medical racism persists. Their manumission upended the conventions of forced labor and confinement, but it opened a new existence, for those who toiled and dreamed.

Born Free

The Canadian scholar Rinaldo Walcott tells us in *The Long Emancipation* about the power of emancipation: "In fact, we must note that at every moment Black peoples have sought, for themselves, to assert what freedom might mean and look like, those desires and acts toward freedom have been violently interdicted." Even when kinships were severed, and bodies were wounded, enslaved people resisted and found a way to find some form of freedom.

Harriet Jacobs's freedom occurred twice; once when she hid in her grandmother's attic, and the second time when she left the US South for the North. The attic shielded her from the potential sexual abuse of the plantation, and Jacobs admits: "I suffered for air even more than for light. But I was not comfortless. I heard the voices of my children. There was joy and there was sadness in the sound. It made my tears flow." But it wasn't just the tittering of her children; it was also the resounding voices of enslaved women who expressed grief when their kin were taken away or the screams of people to public lashings for their attempt to escape. By removing herself from the clamor of her enslaver, Jacobs asserted her political subjectivity while also being reminded of the harm the institution continued to yield.

Slave narratives, like Jacobs's, are not just candid reflections of a writer but deep mental exercises about being free in the world after years of confinement. Jacobs did not hold back; instead, she primed the pages to alert readers about the brutality of America as it was unfolding. The back and forth between revolt and death in Jacobs's narratives was part of the logic of escaping morbidity and confinement which in itself was an epidemic that plagued most Black people in America, while the plantation hardened the body and soul of the captive.

As a historian, one of the great pleasures of my life is reading, but filtering through Harriet's story, and those of other slaves in her time was taxing. I had to immerse myself into a universe that saw Black people as subhumans. But most of all, I thought of the lives of people like Harriet Jacobs, and so many of the millions of slaves and wondered what their lives would have been like if they had been born free.

Nothing could take away the torture and the trauma that people felt on the plantation, or the lives of their grandchildren who might be haunted by being descendants of captives. At the same time, there is a great deal of pride for those who have persevered, and the sacrifice that the enslaved wanted for their descendants, a life where they could be born free. Still, one sure thing is that it was better to die free than to live on a plantation in captivity.

Chapter 2

THE AFRICAN LABORATORY

A scientist in his laboratory is not a mere technician: he is also a child confronting natural phenomena that impress him as though they were fairy tales.

—Marie Curie

The day will come when man will have to fight noise as inexorably as cholera and the plague.

—Robert Koch

THINK OF A LABORATORY. A brightly lit room, bereft of noise and brimming with tenacity. This is a dispensary where adjoining cabinets hold Pyrex glasses, antiseptic reagents, and circular petri dishes. No doubt, there are people pipetting, daydreaming, or scurrying about with half-closed and hastily scribbled-upon notebooks that document the limits of any scientific method and the adjustments they have made to prevail. In the laboratory, rules are established through successful repetition, and celebration comes only when scientists manage to quell curiosity. To work in this setting is to accept that most of the time you will fail, not just because of the lack of skill, but because of the novelty of discovery. Laboratories invite us to translate mystery into certainty, a practice that revels in discovering the secrets of nature, and the origins of color, mired in hours of failed experiments and frustration, with diligence and repetition.

At the turn of the twentieth century, most people thought laboratories could only be found in the grand university halls of London or the natural history museums of Paris. The Polish-born scientist Marie Curie, driven by her carnal desire for accuracy, remarked, "If I see anything vital around me, it is precisely that spirit of adventure, which seems indestructible, and it is akin to curiosity." One of the most innovative and daring scientists of her or any time, Curie constructed her ideal laboratory with enthusiasm and purpose. Her Parisian facility contained conventional beakers, flasks, and test tubes, but also the creative space for like-minded researchers to join in the constant dance between trial and error.

What Curie was able to do was conquer the unknown. Her interest in novelty was not solely a product of these urban spaces; instead, she was part of the arteries of the innovation that flowed between cities and the countryside, along with the single-chambered corridors of a university. Curie, like other scientists, adapted. They were, and continue to be, zones where inquisitiveness is appeased. Under the solemn and massive force of an outbreak, constructing a laboratory reveals far more about the people manufacturing the workshop than the place where it takes effect.

Picture the scene: an elderly scientist with a graying beard, wearing a round hat and shielding his eyes. Dressed in a field suit and boots, he's on one knee with his head slightly bent. Austere and foreboding, his colleague, a beardless staff doctor, gently resting on one knee. Both men appear in the foreground, one sending a sharp gaze at the camera and the other staring at the reptile. The men are encircled by lush vegetation. Behind them two young Black people linger in the monochrome background, perplexed and mesmerized. Barely present, they are subtle, walking alongside a path, neither upset nor elated. It is unclear whether the two European men in the picture were aware of the children in the frame, but the young Africans are there, navigating the rural landscape, living their lives. This is a mise-

Dr. Robert Koch and Medical Officer Dr. Stabsarzt Kleine dissect a dead crocodile which was infected with sleeping sickness. East Africa, 1906. Robert Koch and Stabsarzt Kleine with a dead crocodile, East Africa. Process print after a photograph, 1906/1907. Wellcome Collection. Public Domain Mark. Source: Welcome Collection

en-scène that speaks volumes: European conquest, sacrifice, and apartheid. Both men are hovering above a deceased crocodile, shortly after dissecting the waxen flesh. Their goal was unambiguous, to find the causative agent for sleeping sickness during their research trip in 1906. A closer viewing shows that this cadaverous scene is not a hunt but one in a chorus of efforts to cure an African epidemic.

In March of 1906, when Robert Koch resolved to inspect sleeping sickness in East Africa, he was at the peak of his career, somewhere between an international celebrity and a lapidary civilian. The man next to him was his research assistant, Dr. Friedrich Karl Kleine. But he was not the only one to join Koch on this excursion: Koch's wife Hedwig accompanied him, along with research assistants Max Beck and Arnold Libbertz. Sent by German Emperor Wilhelm II, this small crew gathered at Lake Victoria for a scientific expedition where they established an observation site exposed to the verdant Eastern African landscape.

On his first day on the Ssese Islands in Lake Victoria, Robert Koch asked the German microbiologist Friedrich Karl Kleine: "Can you imagine a better place in the world for working? No disturbances, no visits, and the mail only arrives once in a while!" Writing from the brightly lit countryside, Koch was pleased beyond his expectations. He relished the serenity in East Africa, inspired by the renewed hope and possibility this particular excursion would find a medical intervention for a vicious contagion. Koch and his colleagues glided through the countryside layered in cotton suits, saturated in sweat.

It was common for European biologists to travel to Africa; some did so, as direct colonial agents, and others were part of long-standing science commissions that wanted to observe African epidemics real-time. For the better part of his journey, Koch wanted to replicate his laboratory expertise with care and believed "bacteriology is a very young science—at least, so far as concerns us medical men." He believed he was part of an elite set of people who investigated the natural world, and as such, he executed his travels with stately authority, parched and stern, transposing his laboratory from the stone buildings of Berlin to the open landscape of present-day Tanzania. At the heart of this epidemic was the sleeping sickness, a pathogen that debilitates the body.

Once in a while, I find myself lost in an archive, wondering if I will find anything valuable. So much of the process is built on blind hope. Sitting in a sterile room, in central London, one drizzly afternoon at the Wellcome Collection, my exhaustion was severed by a promising picture—visual evidence of Koch's experiments in East Africa. Most of my finds are physical remnants, scraps of paper, derelict diaries, and photographs that carried the aroma of a mildewed basement. But this picture was different; it was a digital photograph that lacked the scent and present-day archival documents. Instead, the caption confirmed that sleeping sickness was far more pervasive than I thought. The disease not only infected humans but it could be found in wildlife, too.

In 1903, more than two million Africans living near Lake Victoria (present-day Tanzania and Uganda) were infected with human trypanosomiasis, popularly known as sleeping sickness. Gnawed with fatigue, a patient becomes physically and mentally compromised, exhibiting halting tones and trailing off in their speech. The first sign is typical: fever, aching muscles, and inflammation. Perched within the reach of the Horn of Africa, Lake Victoria is the largest tropical lake in the world. The photo of Koch and his colleague is notable, not only for the contrast between the foreground and the background but the integration of the research site in the field. Upon inspection, it reveals a less familiar version of the laboratory photo; however, the entire shows scientists at work.

When Koch went to East Africa, the symptoms of sleeping sickness were well-known. During the initial stage, trypanosomiasis initially lies dormant for several days to a week. With time, a person can develop acute symptoms such as apathy, slow movement, speech disorders, physical weakness, and death. Gnawing at the nervous system, the microbe causes the lymph nodes to swell and the muscles to ache, leaving the sufferer latent. The significant problem for Europeans colonizing Africa was the mass slowdown of African labor and the workers' deaths in the worst scenarios. For Europeans, African lethargy was unavailing, which caused the reduction of their poorly compensated labor. For Africans, the flood of infection was part of the surplus of suffering.

Koch's laboratory in a fecund savannah in one of the minor archipelagos on the northeastern portion of Lake Victoria operated until October 1907. Here, he was allured by nature's marvel, the open sky, in sharp contrast to his Berlin home. Gargantuan crickets sang into the night, echoing their thoughts about the landscape: It was hot and heavy, yet protected from the industrial pollution of central Berlin. Koch was less interested in building relationships with the Luo or Maasai people who populated the land than in the disease that ravaged them. He wasn't there to observe the musical traditions of the region, or to probe into the migration patterns of African pastoralists. What mattered were his experiments.

To Koch, there were plenty of reasons to assay African patients. He and his fellow researchers wanted to observe how sleeping sickness projects and were quickened by the prospect of testing experimental medication on African colonial subjects. After arriving in Muansa, Dr. Oskar Feldman, a physician in the field, described his arrival. "He brought three sick people with him from Bukoma." At the time, Koch developed an untested theory that the tsetse fly was spreading sleeping sickness. Given that the German colony was on the brink of political crisis, Koch made an involuntary stop at Port Florence, and ended up probing the tsetse fly, a potential culprit That July, as rumors of sleeping sickness erupted within the German territories in Mori Bay and Mara Bay outside modern-day Tanzania, the number of cases remained low. Of the two thousand Africans that Koch and Feldman surveyed in Muansa—only forty displayed swollen neck glands, and while many of them were sick with one malady or another, none of them had trypanosomes.

At this point, Koch kept searching for the link between the fly and the disease, but for this to happen, he needed to set up a laboratory in Entebbe for his scientific work. But Koch's account does not end there, nor did he solely rely on the research of other physicians. Koch hoped to create a controlled environment where Africans were "examined, registered, injected, and punctured. The

44

microscopic examinations would take place in two work tents. The location of these buildings is also indicated on the plan." For Koch, he was convinced that launching an on-site camp would be necessary to monitor African patients. Some details were sparse, such as how he would recruit these men and women to participate in the study. Koch's facility wasn't just a benign experience where he could monitor a disease, he was using atoxyl as a treatment for Africans who had sleeping sickness, even though the reagent caused weight loss and blindness. By the early twentieth century, atoxyl had been banned in Berlin, which meant that the only conditions in which it could be administered were outside of Germany and under circumstances where patients were less likely to resist. Indeed, patients who were extremely sick were unable to consent, with a sound mind, to participating in the experiment. Given the toxicity of the drug Koch disseminated to African subjects, he initiated a medical laboratory where researchers would have full access to African sleeping sickness victims.

To Koch's own admission, his experiments were made possible mostly because of German access to its African colonies. As a researcher with the power to knowingly eclipse ethical scrutiny— he could create a laboratory, one that manifested into a medical *Konzentrationslager* (concentration camp). Concentration camps have taken on many forms in German history, freighted by the country's sinister role in immolating and exterminating Holocaust victims. "A concentration camp exists wherever a government holds groups of civilians outside the normal legal process," wrote journalist Andrea Pitzer in *One Long Night*, "sometimes to segregate people considered foreigners or outsiders, sometimes to punish." Although they have been in existence for over a century, concentration camps, as Pitzer argues, have been a way to hold prisoners and other people captive. In general, people held in concentration camps are often separated from civilian populations, and sometimes deemed dangerous to colonial authority, and though some camps involved forced labor, it is not always required.

For East Africans, the medical concentration camps were a form of "disease control," a way to set a group of people apart, devoid of their normal routine, and/or keep their bodies under siege. What makes Koch's concentration camps unique is that they had two purposes: medical experimentation and captivity. African subjects were coerced to participate in regular observation and testing, which were used to advance Europeans' understanding of sleeping sickness. Everything that Koch was able to do, happened under European colonialism of Africa. Unbridled political projects drove a handful of countries—including Germany—to enact commercial, geological, and scientific ventures.

Many long-standing issues with colonial science have to do not with the questions that are asked but how the inquiry is implemented. When Koch arrived in East Africa in 1906, sleeping sickness—and microbes for that matter—was poorly understood, and very little could be done, at the time, to prevent transmission. Even today, the disease still threatens millions of people in sub-Saharan Africa. And while sleeping sickness is endemic to sub-Saharan Africa and was an epidemic in the nineteenth century, according to the World Health Organization, the disease averages several hundred cases per year on the African continent. During the early twentieth century, the British Royal Society funded colonial officers and scientists to inspect and scrutinize sleeping sickness outbreaks throughout East Africa. Europeans—such as Koch—convened in African territories from Sudan to Rhodesia (present-day Zimbabwe) to sharpen their understanding of various pathogens. Still, the difference between Europe and Africa is that East Africans became Europe's research subjects, whether they wanted to or not.

The history of German colonialism in Southwest Africa, East Africa, and West Africa, despite being short-lived, was tied to a political ideology that predicated that some people were civilized humans and others were not. Germans' conception of *Kulturvölker* (cultured people) conferred humanity only to Western Europeans of Christian descent—Africans were conveniently left out. On the one hand, late nineteenth- and early twentieth-century German nationalism was co-constituted with imperialist *Weltpolitik*, a German imperialist policy adopted by German Emperor Wilhelm II, predicated on land expansion. In effect, this narrowed the definition of who could be German, primarily based on phenotype and religion. This perspective matters because Blackness signified "uncivilized" and "foreign" in the German context, a marker that made it perfectly acceptable for them to be politically dominated.

When Europeans invaded the African continent under the guise of scientific research, few questioned why they were there in the first place. Their presence was their mandate. As Helen Tilley notes in *Africa as a Living Laboratory*, "field scientists and their supporters recognized that some kinds of phenomenon could not be investigated or controlled in a confined space." By expanding the physical boundaries of the laboratory outside of a university room, these scientists deemed that their findings, both observation and analysis, would be made more likely to be quotidian than exceptional. But what happens when the laboratory becomes a place of captivity, a morally dubious space? Who decides when the people making decisions about health and treatment do not think to consider the full humanity of those being experimented upon?

The Tsetse Fly

The translucent tsetse fly has an elongated snout, which roils in front of its core and is used to suck morsels of flesh from its prey. Entomologists, who have observed the life cycle of these flies, have noticed that their larval state only lasts a few hours, allowing them to quickly metamorphose from vulnerable juveniles into intrepid adults. Like many flies, they have wide eyes that project forward, with an abdomen that carries the weight of their overlapping wings. Often found in Central, Eastern, and Southern Africa, their power lies in their ability to be discreet: living in placid lakes, in arid terrain, in equatorial forests. They hide in verdant thickets, breed in forests, and linger on buffaloes and elephants. Like hitchhikers who cater to their drivers, they learn to navigate with the aid of others. But this relationship shifts quickly, with the tsetse fly leaving its most dangerous mark in the form of a parasite-induced languor.

Koch, the tsetse fly's investigator, pointed out that some of the sleeping sickness victims had a wave of dizziness; in other cases, they were haunted by their perennial fatigue. Fraught with lethargy, the mind and the flesh are reconfigured, degrading even the most basic actions. But these problems don't stop here; instead, they can persist for several weeks. When left untreated, the disease can cause a coma or death. When death is averted, the damage is mainly in the brain, with the extreme versions of the disease causing irreversible damage. This is where the person begins to change; a patient cannot sleep, initially exhibiting abnormal behavior.

When Koch came to East Africa, he did so with grand ambitions. With an acute awareness of microbial life cycles and a long-held dream, he was part of a fraternity of scientists who had an insatiable pursuit of knowing, dissecting, and exploring. Speaking of his motivations for science, Koch noted in 1908, "If my efforts have led to greater success than usual, this is due, I believe, to the fact that during my wanderings in the field of medicine, I have strayed onto paths where the gold was still lying by the wayside. It takes a little luck to distinguish gold from dross, but that is all." Many elite scientists surmised that their achievements were due to laser-focused discipline and sacrifice. They are not entirely wrong, but the ability to become a scientist in the first place has more to do with one's phalanx of privilege.

Born in Clausthal, a small town in central Germany, Robert Koch was among the premier generation of microbiologists who found themselves consumed with the task of finding the genesis and life cycle of contagion. Science, then, was not treated the way it is in modern times—as a counterweight to more interpretive, less "grounded" fields of study like the visual arts or literature. Koch, rather, saw his vocation as an act of the avant-garde: "No proof had been given that these objects were the cause of the respective diseases, with the exception of a few investigators, who were looked upon as dreamers, people regarded them rather as curiosities than as possible causes of disease." This remained deeply embedded in his practice. Koch examined every microbe and patient with meticulous detail. His methods were unrelenting; as such, he was unmoored by the limitations of early twentieth-century microbiology and put all of his efforts into the craft. For a person deeply embedded in the German Empire, Koch operated, as his colleagues did, using a microscope, traveling, and keeping some financial independence. He could do so, especially given his middle-class background; it was part of the immense power that allowed him to be entangled in a career in medical bacteriology. This sense of wonder saw Koch drawn to the plagues of his time: anthrax, cholera, and tuberculosis. But it also came with his advantage to move throughout the world without much impediment.

Koch often worked in teams with other researchers to try to monitor the pathogens with the tools of his time. Using a microscopic lens and aniline colors as stains, he was able to survey the morphology of microorganisms. Bacteria, it seemed, were not just existing independently of humans, they were wistful agents in our lives. We were co-travelers, sometimes cooperative, sometimes antagonistic, but mostly noncombatant. In 1876, Koch was the first scientist to link a specific bacterium to a particular disease. While this may seem banal today, his discovery sparked the development of germ theory, the scientific thesis that pathogens cause disease. What most of the world understood, at the time, was that bad air, an evil eye, or bad luck were the envoys of maladies. (Some people in the United States still believe this to be true.) Koch's—and other bacteriologists'—digression from popular beliefs did set him apart, and this had everything to do with the mysteries he wanted to decipher.

Like his peers, Koch believed there was more to unravel than the anatomy of the microbe—the environment where it lived had everything to do with whether or not it flourished. So, he photographed anthrax, chronicling the splendor of the microbial world. When I looked at his archives at the State Library in Berlin, I tried to imagine how he transcribed the properties of the microbe to an image, how he negotiated the size of the piece, whether he worked in the plush silence of the laboratory, or if there was doubt. To visualize that which we cannot normally see is how the scientist gestures toward theory. Albert Einstein once remarked, "No one but a theorist believes his theory, everyone puts faith in a laboratory result but the experimenter himself." Looking at Koch's drawings, there was no hint of doubt. Instead, the archives showed the painstaking detail required to document this bacteria's trajectory. The lines were sharp and slight, and carried a life of their own. The image did not crumble on the page, rather it stretched with clear form.

Under the direction of imperial authorities, between 1891 and 1904, Koch's laboratory was based at the Royal Prussian Institute for Infectious Diseases. For years, he worked nonstop, in pursuit of exercising mastery over the microbe. Possessed with ferocious insight, Koch accumulated accolades that were part of the minor signs and symbols of what it meant to be a European scientist at the turn of the century, a growing profession that was international in its approach, and partially aided by the unearned wealth extraction from African and Asian colonies. Koch not only came to help shape the field of bacteriology but also found a place in burgeoning public health institutions. Membership in the Imperial Health Council in Berlin saw him and a global team of scientists develop "Koch's Postulates"—a set of principles that established a causal relationship between a microbe and disease. The principles remain the spawning ground for how researchers conceptualize infectious diseases even today. Though Koch is credited with the theory, the principles reach far beyond one man. Scientists came to their maturity during the late nineteenth century, hoping to accomplish what Louis Pasteur speculated: "Science knows no country, because knowledge belongs to humanity, and is the torch which illuminates the world." This creed was predicated on the proposition that information could be distributed for the greater good and guided many of these scholars, even if, in practice, the method was dictated by a few.

As an unabashed scientist who strolled through German public health and research institutes, Koch was easily recognizable— a bearded, gimlet-eyed man, who was capable of satiating his intellectual queries. Even before he was awarded the Nobel Prize for Physiology/Medicine in 1905, he was obsessed with outbreaks. In a

1904 letter to the Minister of Medical Affairs, Koch agonized over a budding surge of sleeping sickness, citing an 1896 epidemic on the north shore of Lake Victoria, which was believed to have infected up to 200,000 in East Africa. After reading a report by Dr. Feldman, which was published in October of 1903, Koch believed sleeping sickness posed a threat in German East African colonies. Koch was prudent, feeling apt to suggest a course of action—that the German government establish a scientific mission. In his memorandum, Koch refers again to British, French, and Portuguese scientific missions being more successful in their management of East African epidemics, asserting in July 1904 that "German science should not be left behind in this respect."

On the surface, the scientific expedition was about solving sleeping sickness, but it was ultimately about maintaining German imperial power, partially because the risk of spreading the disease throughout the colony could pose a threat to the extraction of natural resources or transporting for export—mostly done by Africans. The power of European colonial bodies to mitigate health risks was not just a matter of establishing an immunization campaign, their efforts were bound to use parts of the region for military purposes. For the most part, the physicians that were recommended to join Koch were part of the *Schutztruppe* (protection force), a German colonial military unit that was part of the colonial units. By September 1904, the president of the imperial health office approved Koch's request, arguing "there is no doubt that the spread of the disease to our protected area [German East Africa] is only a matter of time," fearing that the disease was both dangerous and likely to spread over Africa.

Plenty of funds were distributed to scientists, but administrative and military divisions received far more assets. While the imperial office dedicated 10,000 Marks for the scientific equipment, they allocated 50,000 Marks for other costs. The German health initiative could never be fully divorced from their imperial efforts, whether or not Koch—or any other scientist—was motivated to cure sleeping sickness. Incidentally, he had the enviable quality of a scientist that was driven by zeal. For Koch, the microbe was an invasive species that could transcend physical or political borders. Given that the German East African colony (present-day Tanzania) shared porous borders with other African colonies, Koch believed that the German Empire had to act.

The more invasive he perceived sleeping sickness to be, the more he bought into its menace, and the more pedantic he was in his methods. Out in the field, Koch offered a clear answer: Tsetse flies were the vestors of contagion. But knowing the cause wasn't enough, he wanted to answer a more pressing question: "How can you treat the ill?"

Stationed at Lake Victoria, in tents, Koch and his research unit turned to the preceding work of other scientists to determine which drugs to use, a task that required human subjects. As the team would learn, there are no straightforward experiments in the field. The late German historian Wolfgang Eckart, who wrote about sleeping sickness in German East Africa, notes, "After the Reich Health Council had passed regulations in line with Koch's recommendation [to create sleeping sickness camps], the colonial government in German East Africa immediately set out to apply them." While the number of East African patients recorded was approximately twelve hundred people, the actual number is unknown.

And that need—to Germany's eye—was supported by the country's feelings on its colonial holdings at the time. Germany left no ambiguity—its colonies were ruled with stern and direct force. That extended to its extraction of resources, colonial governance, and their convictions that Africans were the most appropriate experimental group. For most Germans, this history is largely concealed, a reminder that not all histories laid equally bare. .

Konzentrationslager

As early as ten years of age, well before I moved to Germany in 2017, I was taught about the trains that transported Jewish families from their homes to the Nazi regime's ghastly death camps. And later, I would learn that people who were branded with the pink triangle—an emblem for queer people—were also massacred. I learned about these abuses of power and degradation of human dignity through a wellspring of sources—my history textbooks, popular films, and more tactilely, museums. These accounts passed on in the United States were a critical reminder of the cruelty of men, reminders of a history affirmed by countless scholars in the hopes it shall never be repeated.

Those lessons, however, never extended to the concentration camps established in Africa—at least, not until I moved to Germany. When most people think about "concentration camp" the term immediately evokes "Nazi extermination camp," but the practice and technology is older and far wider reaching than what I could ever have known.

As early as 1899, during the early part of the Anglo-Boer War (1899-1902) in present-day South Africa, the British government established concentration camps to target Boer (European) settlers and anti-colonial Africans. Tens of thousands of Boers and Africans died in the overcrowded and unhygienic camps from malnutrition, dysentery, typhoid fever. Between 1904 and 1908, the German Empire used a similar military tactic against the Herero and Namaqua dissidents (present-day Namibia). In this case, several thousands of women, children, and men were placed in concentration camps, leading to a genocide of these two African ethnic groups. For civilians and rebels that interfered with colonial expansion, African concentration camps became a swift and cruel method to eliminate insurgents.

Addressing the imbalance in how we discuss these historical events is an enormous undertaking and is not merely a prescriptive measure but a reflection on the moral function of history. The question of how this aspect of the past could go so unknown to my peers in the United States irked me and still does. My queries were counseled by theorists, joined by a long-standing recognition of what some Africans already knew—Europe not only pilfered Africa and allowed its people to lie fallow in the ways most everyone has been told, but it did so also through its unsanctioned and circumscribed medical camps. "Colonial conquests constituted a privileged field of experimentation," wrote the philosopher Achille Mbembe in his book, *Necropolitics.* "They gave rise to a thinking about power and technology that . . . paved the way for concentration camps and modern genocidal ideologies." Painfully aware of how colonialism instantiated scientific experiments, Mbembe sheds light on the damage that was done when Africans became nonconsensual research subjects. Even if it might be rendered somewhat innocuous by our awareness of the scope of inhumane conquest across the African continent, the colonialism of Africa was a testing ground for medicine, but also for the fascist rule as we know it.

To be clear, the political concentration camps installed in German Southwest Africa gravely differed from the medical concentration camp. It's important to understand the seminal influence of the science developed on unwilling, if not enslaved, Africans in the earliest days of modern medicine. This science was developed without care for well-being, without acknowledgment of its test subjects' suffering, and with no remuneration for the work and its results—undoubtedly blood extracted from the sick, sourced straight from the blood and bodies of colonized Africans.

Even Koch observes that some of the patients at one of his East African medical concentration camps were not interested in the business of being held captive. Devoid of rancor, speaking of patients held in medical camps near Lake Victoria, he wrote: "As soon as they feel better or the cure becomes long and tedious, they break it off and run away. Many of our patients, who had come from abroad, also had to stop early because they could no longer leave their property unattended or ran out of money to support themselves." The statement reveals that, even when some Africans initially yearned to seek treatment, the drugs and detention meant that many of them wanted to escape. Reading this in the archives, I was astounded by Koch's honesty, easily distinguishable from the remainder of the text, a localized scientific report about colonial East Africa.

It is even more disturbing to digest how occupation in this scientific enterprise was made legal, namely through ministerial orders. European scientists were able to conduct their research, without significant oversight from civil and political administration. This was especially true for European scientists who were part of sleeping sickness commissions. When Koch went to East Africa, the British Society for Sleeping Sickness established a scientific mission to understand the nature and method of transmission and the most suitable treatment. For most people in the early twentieth century, the knowledge of the connection between a microbe and a disease wasn't prevalent. Nevertheless, Koch's expertise and acclaim meant that colonial administrators believed microbiologists—like himself—could solve colonial medical problems. Germ theory, in effect, became a way to develop a pedantic and lucid criterion for contagion. It required strict surveillance, under the guise that human experimentation was necessary. Although coercion and harm were present, Koch's *Konzentrationslager* (concentration camp) took on a different form than the concentration camps that were established in German Southwest Africa (present-day Namibia) or Central Europe.

Very quickly, Koch installed his lab in the field. Within a month after arriving in East Africa, Koch erected the Bugala sleeping sickness research camp, one of the most significant sites for his study. The ill were collected and confined to the camp, where their tents clammily stifled their mobility. I imagine that when they were sitting in their tents, they hoped the world would pass faster or that their pain would go away. Not all of the sleeping sickness patients entered by force. Koch's records show that about half of the Africans at the camp initially sought it out in search of remedy Some Africans came from the main island of the Ssese archipelago; with a little more than two hundred other patients from the other surrounding regions. At times, people seeking medication lived adjacent to the camp, in other cases people began to create their own shelters near the Bugala campsite—one of the main sleeping sickness research facilities. Africans built a double barracks, constructed with a wooden framework, enclosed with grass walls and a grass roof. The huts, according to Koch, were intended for family members who brought their relatives to these medical sites, hoping that their loved ones would receive therapy. "A village has been created next to our camp," Koch notes.

When I first saw photographs of these camps in the archives, beaming patients was not what I saw. Infected people were put into isolation and monitored by scientists, who incorrectly believed that their internment would protect the rest of the African population through their isolation. Over and over, Africans were placed into these camps, usually one thousand at a time, unable to fully possess their land and space.

While at Bugala camp, African subjects would have their eyes, ears, and limbs examined, with doctors eventually puncturing them with a needle to extract what they referred to as *Krankenmaterial* (sick material) from their bodies. Like many scientists jolted by empiricism, Koch organized his subjects with visual markers. Those housed at the Bugala campsite had wooden identification tags attached to their wrist or hanging around their neck. Several times a day, the researchers approached the patients, searching for signs of malady by examining the pupils' dilation or the neck's swelling. Throughout Koch's archive, he indicated the following:

No. 168 (Bugalla) A. Woman of 24 years. Sick for 3 years. Admitted on 1 October. She is so weak that she is led or rather half carried by her husband. Severe drowsiness. Wets herself. At times subnormal temperature (prognostically a very bad sign). Now she goes without help, does not wet herself anymore. Temperature normal. Drowsiness has diminished. Mentally still somewhat dull, but in steady further improvement.

No. 236 (Bugalla) T. Man of 30 years. Catechist of the French mission. Sick for 2 years; has been unable to walk for 6 months, has been in a sleeping state for 3 months. When admitted on 11 September, he was quite helpless and weakwilled. He constantly laid in deepest sleep and wet himself. Aroused, he opened his eyes blinking for a few minutes, yawned continuously and then fell asleep again. Now he has completely lost the hypersomnia and with it the enuresis. He is fully conscious, can walk well, even goes for walks alone.

He speaks quite intelligibly and can read aloud from a book. The improvement is still in progress.

No. 527 (Bugalla) D. Man of 32 years. Sick for 2 years. At admission on October 15, very weak, so that he is unable to walk. Subnormal temperature. Pulse very frequent and hardly palpable. Wet himself for 3 months. Heavily dazed and almost continuously asleep. Even now he sleeps a lot but no longer wets himself. Can walk when supported by only one person, whereas before it required two people. Mentally freer. Pulse slow and vigorous. Temperature normal.

Koch's notes acutely document the demographics and physical attributes of the patients—their age, gender, and symptoms. But nothing about who they are, their names, their wants, and their ethnic groups. He is inclined to highlight the African subjects who were successful, the ones who were able to recover from their slump, even if they were still recuperating. What it suggests is that the relationship between a scientist and a research subject can be undeniably charged, an uneasy mix of dominance and subordination, but something else emerges in a colonial context like this one. As we move between his accounts and read through his thick description of their conditions, their foggy memories, and even their unrestrained bowels, we realize that there is something lacking in care throughout his frank remarks.

As historian Manuela Bauche has written in her article "Robert Koch Sleeping Sickness and Human Experimentation in Colonial East Africa," it is unclear if the African research subjects volunteered and whether they knew the risks of the experiments. Bauche diplomatically expresses one of Koch's flaws: "Koch's main task was to test suitable remedies against infection and to develop an effective method of application. However, no therapy against sleeping sickness had yet proven itself." Koch's lone "remedy" was to inject African patients with atoxyl, a novel arsenic compound that was even then known to have noxious properties in human subjects. Even the patients in whom he observed progress did not benefit for long. Koch wrote, "The improvement in the condition of the patients made further progress in the beginning, but then came to a halt after a few weeks." Some who received treatment had an impassioned daze after spending weeks being prodded, and atoxyl could also lead to blindness if administered for long enough. In some cases, Koch recognized that "the subcutaneous application of these drugs [atoxyl] is quite painful," but his report makes no mention of measures taken to alleviate the pain of his African patients. He goes on to admit that while the drug showed promise in nonhuman animals, the results for humans were far less clear.

The French novelist Marcel Proust once wrote, "We can find everything in our memory, like a dispensary or chemical laboratory in which chance steers our hand sometimes to a soothing drug and sometimes to a dangerous poison." In Koch's atoxyl-laden camp, there were 1,633 patients, of whom 131 died over the course of the first ten months. He notes that some of these patients were seriously ill, and others responded poorly to the reagents. In some cases, there were patients whose conditions deteriorated. "In total, we observed 22 cases of blindness," Koch writes on April 25, 1907, most likely caused by the atoxyl treatment. Koch's notes on treatment are prosaic. He is dispassionate about African suffering and death. What seemed to matter more was the art of constructing a quarantine site, documenting and charting an experiment, and creating conditions where disease progression could be readily observed in human subjects. This was a research opportunity first, a hub for innovation rather than a humane campaign.

Despite its harmful side effects, the misstep in giving Africans atoxyl calls into question if Koch wanted to cure sleeping sickness. The sleeping sickness commission's intention was to treat the disease, but in practice, it restricted freedom. Experimentation is a delicate process; one has to design, enact, and repeat the methods. When prompted by the German imperial government, Koch demanded quarantine internationally: "Traffic restrictions, border closures, and international agreements are necessary to prevent the disease introduction from other areas." This opens up a set of questions about whether his aims were primarily concerned with healing the sick or inhumanely using colonial relations to experiment on Africans; could their illness or perceived illness be justification for preventing them from traveling?

Science's ambitions have been resolute about exercising prudence. Aristotle once remarked, "Conscientious and careful physicians allocate causes of diseases to natural laws, while the ablest scientists go back to medicine for their first principles." A polymath philosopher, Aristotle created an aphorism stacked with a moral premise, medicine's purpose is to alleviate harm. Against the backdrop of miasma theory and Galenic treatment, he was advocating for a world where physicians could, in principle and practice, cure the ill, extracting a sickening ether or texture from the infirm. What was missing from Koch's experiments were vital details, the way the experiments came to be, and how the worship of science turned African people into dark matter.

Koch's scientific expedition to East Africa, even for his time period, raises ethical concerns, not only because he was aware of the potential impairments that could be caused by atoxyl, but because it reveals the brutishness of colonial medicine and begs the question if European scientists could see Africans as full humans. In *Der moralische Diskurs über das medizinische Menschenexperiment im 19. Jahrhundert* (The moral discourse about medical human experimentation in the 19th century), the German historian Barbara Elkeles investigates the ethics of medical experimentation, drawing attention to Koch's ethics. What she notes is that during the 1890s, he actively promoted the use of tuberculin, a compound that he purported could treat tuberculosis, even though it was therapeutically ineffective. She evinces that the undertones of human experimentation were not solely tied to benevolent notions of scientific progress; instead, they were sometimes linked to ego and ill will. For late nineteenth-century scientists such as Koch, the Aristotelian principle to heal was occasionally nonexistent, even in some European human experiments, but the ethics of care were even more absent in a colonial context.

In his seminal book *How Europe Underdeveloped Africa*, the late Guyanese historian Walter Rodney provides receipts of the damning violence white people imposed on their colonial subjects, asserting that "North Africa and the Sahara became available as a laboratory for the evolution of techniques of armored warfare," and "Ethiopians were used as guinea pigs, upon whom Italian fascists experimented with poison gas." Rodney's text documents the steady flow of resources, labor, and humans that Europeans extracted from Africa, and it does so by diagnosing the institutions that made it possible. It is unsurprising to hear that colonialism extirpated wealth from Africa, but there is still something troubling that twentieth-century experiments on Africans are viewed with little ethical compassion by most Europeans.

Never mind that Africans had their blood extracted and injected with a poisonous compound; the laboratory was more a site of microbial conquest within the colonial ecosystem. Laced with conviction, there are moments when Koch's work inspires the reader, the willingness to understand the interconnectedness of the human body, forming solidarity with the microbes, while at the same time wanting to concur with it. Koch appears to exercise authority over several camps primarily because he was a world-renowned scientist and part of the upper echelon of the colonial hierarchy. By the beginning of 1907, the health of Koch and his wife began to take a different turn. From the outset, Robert and his wife Hedwig could live freely in East Africa, though over time, they experienced a mixture of nagging troubles—an illness of their own. In 1907, Hedwig Koch left East Africa after suffering from a malaria infection. Shortly thereafter, Robert contracted lymphangitis, a malaise that emerged from sand flea bites. In his journal, he described how, for weeks, he was "tied to his hut and could hardly limp over to the lab tent." But as the experiments continued, his plan was carefully wrought, especially when a new layer of German colonial scientists built upon and expanded Koch's medical policies.

When Koch left for Germany in October 1907, he transferred his responsibilities to other German researchers. His presence in East Africa was part of a broader effort by European governments to situate science in the African colony against widespread efforts by African research subjects. Two months later, the Imperial Health Office formally designated Dr. Kleine as Koch's successor and director of the sleeping sickness campaign in German East Africa. Having previously served as a medical assistant to Koch, he was attuned to the research site and subjects. Koch's influence did not stop with designating a replacement; he found a way to insert himself in colonial policy, offering nontherapeutic measures for addressing sleeping sickness. On November 18, 1907, writing to the Reich Health Council, Koch delved into an essential quarantine measure: transporting infected districts to noninfested regions. He believed that mortality was inevitable without treatment and that infected individuals would, without exception, die. Even when his recommendations acknowledged an insect was likely responsible for spreading disease—which was difficult to sequester in nature—Koch recommended that colonial officials establish encampments to house sleeping sickness patients.

Koch's recommendations included changing the African landscape. Koch's counsel for sleeping sickness was wide-ranging: He endorsed deforestation, stripping swaths of land of their foliage in the hopes that the vector of sleeping sickness—the tsetse fly—could be eliminated if its food source, namely infected people, were isolated. "The camp must be set up not too far from inhabited places, especially where there are no tsetse flies," wrote Koch in his 1907 essay. He admits that there were inconsistencies in some of the data, but the medical concentration camp became central guidance for the German government. After the Imperial Health Office passed regulations in line with Koch's counsel, the colonial government in German East Africa immediately set out to construct more camps in the image of Koch's first. Three more sleeping sickness camps were established in the territory, two in Lake Victoria (at Kigarama and Schirati) and one more at Lake Tanganyika. More than 1,200 patients were isolated and treated, though not very successfully. Isolation, in this case, was not merely about separating Africans from each other but about creating conditions so that the tsetse fly could not escape and survive to take root elsewhere.

Over and over again, with the recommendations Koch provided, familiar experiments happened furtively, in the field, in open spaces, in a tent, for some to see, but mostly on those in captivity. It becomes clear that the concentration camps' goal was to test many people with a reagent under the guise of scientific research.

Between 1908 and 1911, the ten sleeping sickness camps administered by Dr. Kleine and other European military doctors in German East Africa showed very little progress in curing the infirm. Paul Ehrlich, a German physician and researcher who conducted research in Africa shortly after Koch left, developed another arsenic-based compound, arsenophenylglycin, which is commonly used to treat syphilis, which Ehrlich administered injections of subcutaneously to African subjects. In some cases, the physical impairment from sleeping sickness could not be reversed with any treatment. On the surface, the drugs—if they worked—could minimize traces of the protozoa that caused sleeping sickness, but if they did not, the arsenic within most likely would gurgle and run through the organs, destroying the intricate flora and fauna of the abdomen.

Throughout the German colonial period, these experimental drugs posed harm. Another physician, Dr. Scherschmidt, based at the camp of Utegi (present-day Tanzania), carried out an experiment where he administered arsenophenylglycin to thirty-five patients. Fifteen of them died, with another six left physically impaired. He cited that this "resulted from intoxication." As German colonial science gathered force, African colonial subjects continued to challenge the experiments with disparate acts of rebellion. For example, three of Scherschmidt's patients managed to escape, evidence that not everyone "treated" by these medical camps had done so willingly. For Koch's successors, the ineffective and reluctant patients foiled the experiments.

Pulsing beneath the mass resignation were the relentless efforts by the Africans to free themselves from the scrutiny of observation and regular injections in the bivouac. Over one thousand patients left the German-led medical camps, their refusal a testament to their agency. When Scherschmidt returned to Germany in 1911, the Imperial Health Council decided to ban the use of arsenophenylglycin, noting the reagent was ineffective. After 1911, most of the German East African colonial medical camps and posts were abandoned. Some colonial officials surmised that if they could get rid of the savannah woodlands, then the tsetse fly and, by extension, the parasite which causes sleeping sickness, would no longer pose a threat.

Sleeping sickness missions were not the only scientific endeavors carried out during the German colonial period. German protectorates in Cameroon, Togo, East Africa, and Southwest Africa gave physicians, biologists, and agronomists liberal access to African land and its people. It was not uncommon for German officials, traders, and settlers to be accompanied by government-trained colonial doctors to protect the health of colonizers. Rather than focus on a single illness such as sleeping sickness, German scientists and their counterparts sought to expand their scope by studying a host of maladies, under the rubric of tropical diseases, such as malaria and dysentery. Sometimes, German field doctors also forged connections with other European colleagues, especially along the intersections of colonialism and tropical medicine.

While Koch's colonial reputation is often perceived as ambiguous, some of his disciples were far from neutral in their methods. One of his admirers, Claus Schilling—a German physician who was an administrator in colonial East Africa—actively carried out inhumane experiments in Africa and later under the Nazi regime. A 1910 essay, "What Is the Significance of the New Advances in Tropical Hygiene for Our Colonies?" lends full-throated support of German imperialism. Awash with colonial fervor and working closely with French, Belgian, and British counterparts until World War One, Schilling believed that carrying out experiments, even with human sacrifice, was essential to finding a treatment for infectious diseases. From his makeshift laboratory, he presided over the colonial African malaria commission, playing a prominent role in documenting the medical landscape of the overseas German colonial states. These experiences informed a principle that was the core of the experiments: Europeans should exercise their dominance over Africans.

Schilling was not unique. Hans Ziemann, a German colonial doctor, alleged these medical trials had practical benefits for African people and that this intervention would "improve the population socially and hygienically." In 1903, when he joined the *Schutztruppe* (protection force), a German military unit, Ziemann was also head of the civil and military medical services in Cameroon and conducted drug therapy experimentation, along with the intensive surveillance, forcible confinement, and quarantine of Africans. Later, Ziemann published material that argued for racial apartheid: the separation of European residents and African domiciles. His justification was the high malaria rate among the African population. He did not account for how poor housing and exploitative labor, constructed under colonial rule, contributed to disproportionate rates of infection for African communities. Nevertheless, Ziemann's career flourished, and he pursued publishing opportunities in the *Journal of Tropical Medicine* and was well-known on the European scientific conference circuit. In 1938 he became head of the tropical medicine–parasitological department of the Military Medical Academy in Berlin, which he co-founded. Schilling and Ziemann were not rogue doctors who conducted experiments on Africans; they were part of a colonial system that rewarded Germans who experimented on Africans.

Race Science

As director of the Kaiser Wilhelm Institute of Anthropology, German social scientist Eugen Fischer participated in the campaign to measure and catalog the skulls and physical features of African prisoners in Southwest Africa, establishing a field of "race science" he hoped would prove the alleged superiority of white Europeans. His colonial interest in Africa helped him climb through the fascist regime in Nazi Germany, where he later served as a "racial hygienist" under Adolf Hitler's reign. Similarly, the German zoologist Leonard Schultze experimented on African bodies, acquiring approximately three hundred skulls of deceased Nama and Herero prisoners from German Southwest Africa concentration camps. These two men were part of a network of scientists, officials, and administrators who promoted the racist ideology of the Nazi regime.

There was an affinity of sorts between the nineteenth-century scientific aspiration to design a controlled lab environment and the corresponding nineteenth-century military inclination to create a controlled population. African bodies had been made disposable, precisely because they were colonial subjects who were relegated to a separate set of rules from Europeans. The instinct toward domination and subjugation, utilizing humans as playthings, did not stop with medical racism or pseudoscientific ideology; it proceeded in the form of the planned massacre, with German scientists serving as arbiters and executors of German colonial science. The systematic forced removal and dispossession of Africans, through the militarization of Western, Eastern, and Southern Africa, was paramount to their premature death during German colonialism. A regime predicated on seeing Black people as immune to pain. Given that he was a well-decorated international scientist with less of an imperial footprint, Koch's colonial imprint does not stain like Schilling's, Ziemann's or Fischer's. Instead he is globally renowned.

One can still find Koch's fingerprints on a host of contemporary biological topics: the origins of contagion, the specifics of the cholera epidemic, and the broader conception of the corrosive effects of epidemics on people's bodies. His research, writing, and criticism brought an incisive and prescient insight on the life of microorganisms at least in some part because he was free to learn, travel, and document the molecular world on his own, now intolerable, terms. Koch was not the first nor the last to orchestrate an experiment on

Africans, but the legacy he left was deeply embroiled in medical camps that made it acceptable to experiment on African bodies. For most of the twentieth century, his legacy was well received. However, that began to change in the twenty-first century, when Germans were confronted with a new epidemic.

The Legacy

During the height of the Covid-19 pandemic, when the world was at a standstill, I was glued to my laptop, reading public health data from any institution that could help visualize the virus. Perhaps ironically, The Robert Koch Institute, Germany's premier public health institute, became my primary guide. I religiously checked the institute's website to find the Berlin Covid-19 infection and death rates. I was informed about the raw data and learned about the government's regulations on social distancing, group gatherings, and masks. My obsession with visiting the Robert Koch Institute's website was part of a habit, a routine that I and many others I knew developed to protect us and to stave off Covid-19. And as a former biologist, I trusted scientists in Germany (and elsewhere) to report on the conditions of the disease. Just as I became aware of this new era of living in the age of coronavirus, I was also made aware of how racial and, by extension, colonial violence in Africa shaped social relations today.

After the murder of George Floyd in the late spring of 2020, anti-racist activists in the United States called into question the merchants, administrators, and governments that had profited from the oppression of Black people. The year 2020 not only marked a broader conversation about racist trends in the United States but also probed people in Germany to reconsider the legacy of their saints. We still grapple with the legacy of colonial violence in former African colonies, recognizing the question of historical amends raised by this past. As Angela Saini argues in *Superior: The Return of Race Science*, modern biology emerged in the name of empire with the aim of promoting European white supremacy. Tragedy opens the floodgates for anti-racists in Germany to question this history. But consideration and reflection aren't enough, especially when economic and power imbalances created during the colonial era still persist.

Modern society holds up in order to convince itself that all is well: comforting fictions that rest upon amnesia and ignorance, not only about its past but its present. Still, the Covid-19 pandemic, and the Black Lives Matter protests, in part, opened up vital debates about the origins and legacies of our esteemed historical figures. In the United States, Confederate artifacts received fresh rounds of criticism. In Germany, where I lived, activists called for removing Robert Koch from the Institute of Public Health in light of new attention on the man's colonial activities. In my own re-evaluation of Koch, I found multiple statues and plaques commemorating the man and his work.

During the late spring of 2022, I explored the site of his former laboratory, now a square bearing his name and adorned by his statue. The gray summer sky stood still against the backdrop of the hustle of Berlin Mitte, a borough whose name means "center." Blanketed by the neighborhood's silence I paused, staring at the larger-than-life figure. One of the things that makes a person a legacy is their ability to have their memory live on well beyond them; it comes with a series of questions: What would the world look like if not for that person's invention, poem, or building? I kept wondering, how do we remember scientists with a complicated past? For many contemporary scientists—German and non-German—Koch is a demi-god, a symbol of pride, who should be devoid of criticism. The scientist is valorized from the start, producing an atmosphere where they and their principles become part of a hagiography. This deal is part of the incipient war between, in this case, a scientist's legacy and African bodily autonomy, a reality that diminishes the suffering of those in the past.

The answers to these questions are not straightforward, nor is the consensus about Koch. In pursuit of this history, I turned to Professor Thamil Ananthavinayagan, a legal scholar born and raised in Germany to parents from Sri Lanka. As he saw it, Koch had inhumanely led a campaign to cure sleeping sickness several times, allying himself with other physicians and researchers, but more explicitly, the German colonial state. "Experimentation was part of the colonial project. And in many ways, it was reproducing itself," he told me. "I think the Germans, the German culture, choose collective amnesia." As he reflected on the early days of Covid-19, he remarked on how French medics and medical scientists proposed to test the early stages of the vaccines on sub-Saharan African people. For Professor Ananthavinayagan, "to accept Koch's view in a positive light, is a reproduction of the colonial mindset of the past."

The medical concentration camps in Africa were both the definition of the European colonial establishment and the personification of the flexibility of the African laboratory. Koch's career, launched in Germany and carried out in a European colonial outpost, offers insight into medical experiments under conditions of confinement. The laboratory has, in many ways, opened up the possibilities of healing, giving us an entry point into the cell's life cycle, the cause of contagion. Yet the questions of how we intervene to affect the course of an illness or whether experiments are done to learn new information are predicated on the ethical pillars of our society. A core value of science is that experiments are not merely a means to an end but also a reflection of whom we consider human and worthy of autonomy and movement.

Many people want to trust science and scientists to make decisions predicated on reason. We hope that the questions start from a place of the common good. But the historical wounds linger consciously and subconsciously for many people of African descent.

Black people's concerns, their humanity, and their trauma ought to be given due respect. European colonialism on the African continent's past is not merely reflected in the objects or people it stole or the wealth taken, but in how anti-Black ideology continues to operationalize difference. It lives on in how police forces are often trained to target people with migrant backgrounds, contributing to racism and discrimination. And it also resides on whether we choose to evaluate critically the ethically dubious experiments in colonial East Africa that undergirded and were later directed toward the Nazi medical regime.

There is no reason to believe that such progress couldn't have been made in humane conditions. What would have happened if, in 1906, Africans had been allowed to cure their own ill? This question is particularly significant as we continue to understand Europe's relationship to the African continent. I wonder what the world would have looked like if Africans in 1906 had access to the tools to carry this out. Wealth, power, technology, and training could have been distributed without the violence and inequality Europeans used to amass them.

What if, instead of Germany capitalizing on this crisis to strengthen its colonial rule and, by extension, even further subjugating Africans who saw their lives treated as insignificant, they had legitimately worked to empower the people to foster domestic care for the sick? This is what a true partner or ally does. Instead, in this case, we've seen the overseer of immense death and destruction treated as a hero back in his home because his acts perhaps lead to some good long afterward.

This is the lesson of Robert Koch's Africa.

Chapter 3

WHO'S AFRAID OF THE FLU?

I had a little bird
Its name was Enza
I opened the window,
And in-flu-enza.
—1918 children's rhyme

It was a web, a net, spreading wide and enmeshing every sort of
cousin and dependant and old retainer.
—Virginia Woolf ("On Being Ill," 326 version)

DURING THE EARLY MORNING OF March 4, 1918, three men
in Kansas dipped into an acute fever, forcing them to step
away from their military duties. Stationed at Camp Funston, private Albert Gitchell, Corporal Lee W. Drake, and Sergeant Adolph Hurby endured a deluge of symptoms: a roaring headache, a suffocating sore throat, and extreme muscle pain. With hollow cheeks, their bodies were soaked with perspiration, an effect meant to cool their shivering bodies, as they slowly fought a pathogen. Gitchell, an army cook, alongside other chefs, nourished thousands of the soldiers at Funston, while Drake and Hurby prepared a young generation of men who were drafted into the US military. Camp Funston, a training ground for fifty thousand soldiers and a detention center for pacifists who refused to fight, sat on an alluvial plain adjacent to the Kansas River in the town of Fort Riley.

Far remote from the bustling ports of the northeastern corridor, the sluggish creeks passed through this midwestern town, where American men deployed from Illinois and Texas were transferred to prepare themselves for battle in the First World War.

When the three men admitted themselves to the camp's infirmary, medical staff sprang into action, attending to their needs by measuring their temperature and monitoring their affliction. Once the camp medical officer confirmed that these men had influenza, they were sequestered. Later that day, however, over a hundred men pulled themselves into the military's clinic, seeking reprieve from the flu virus. By the end of the month, Camp Funston established an emergency hospital for a thousand of the army's servicemen, hoping that the disease would settle. Whether they were wholly aware or not, the men at this camp were harbingers of the 1918–1919 flu pandemic that swiftly affected many of the people at the base and across the world.

By the end of March 1918, over six hundred of the men at Camp Funston became ill with influenza, with many of them incapacitated, unable to carry out their service duties. Like Gitchell, Drake, and Hurby, many of them felt wretched from fever, muscle pain, and inflammation. As the week carried on, flu patients could experience sore throat, a runny nose, and the loss of appetite. Some men complained of abdominal pain, and others vomited uncontrollably. Other soldiers lay stricken, with their faces sullen, their airways obstructed by their mucous membranes." As the number of indisposed men grew, the military set up a communal sickroom, an infirmary that had been converted into a makeshift hospital where young soldiers lay near-comatose on angular cots, wrapped in pillows and linens. Staff began to don protective gauze face masks, as the recently able-bodied young men languished, incapacitated not by the expected war that had brought them to the camp, but by an invisible predator in their midst.

If photographs sharpen our understanding of the past, the image of the emergency hospital in Camp Funston unearths the unsettling consequences of the 1918–1919 flu pandemic. What seems like a grainy and visually dull picture brushes against the gravity of the disease. For the military men, the outbreak wasn't merely a solitary illness, where they could recover in silence; it was visibly charged; a collective state of agony. When a US military officer took a profile picture of these men in their sickbed, it affirmed that even the strongest men could be subdued by a virus.

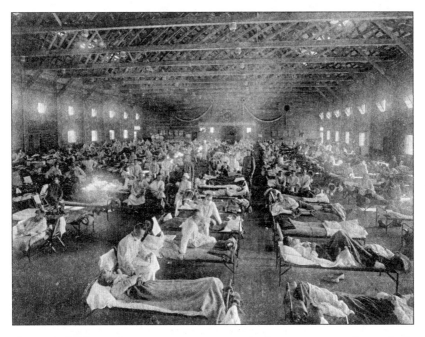

Photograph of military patients at Camp Funston, Kansas, taken by anonymous army photographer in 1918. National Museum of Health and Medicine

Although the image was anonymous (the photograph is credited to an unlisted military officer), the rows of beds expressed the scale of the outbreak and converted the private consequences of mass outbreaks into a public spectacle.

Camp Funston was the first, but not the only, US military base that dealt with the flu pandemic. In September 1918, approximately ten thousand soldiers at Fort Devens—near Boston—contracted the virus, resulting in dozens of deaths. Built in 1917, Camp Devens, like Funston, served as a training ground for drafted soldiers who were funneled into the First World War. By the end of the month, over 10,000 of the 50,000 soldiers reported flu-like symptoms. A doctor who treated influenza patients wrote to his friend about the outbreak: "This epidemic started about four weeks ago and has developed so rapidly that the camp is demoralized and all ordinary work is held up till it has passed." Despair was rampant, but it was not the only response by the public. His testimony was neither original nor profound, but it spoke to the confusion and terror that became part of civil society.

Later, between 1918 and 1919, fifty to one hundred million people died of a new, even more deadly strain of influenza—some reportedly within hours of contracting the illness, others within days. The socioeconomic fallout was immense: garbage collection postponed, farm workers forced to delay their harvests, and businesses forced to close their doors. Given the loss of labor from war and disease, the United States experienced a mild economic recession during this period.

Frontline physicians toed the line between resilience and abrasion, sometimes unable to provide optimal care. Not only were health workers at exceptional risk, but they were also overworked. A December 1918 article in the *Bulletin of Woman's Medical College of Pennsylvania* noted: "If the work at the main hospital was strenuous and heart-breaking, the work at the 'front,' meaning Front and Ellsworth Streets, where our emergency ward was established, brought us very close to the realities of war." Founded in central Philadelphia in 1850, the Woman's Medical College was the second oldest medical institution in the United States that trained women in medicine. As the historian Ellen S. More notes in her article, "A Certain Restless Ambition," although few in number, American women physicians who served on the frontlines of World War One attempted to assimilate into their professions by exercising efficiency in care. On the one hand, that meant fighting a war but on the other hand, that included documenting how the influenza outbreak—at home and abroad—left people forlorn and debilitated.

By October of 1918, the United States began to feel influenza's wrath. That winter, officials noted to the public how the ailment "spreads rapidly where people are crowded together in railway trains, in theatres and places of amusement, in stores and factories and schools." By reporting this information in the *New York Times*, there was recognition that the media provided a powerful tool to put public health into practice. Despite these efforts, for most Americans, influenza was truly indiscernible, misunderstood, irrevocably infectious, with little treatment for the frail. What was even more difficult is that the very spaces where people gathered— churches, mills, and restaurants—were where the virus continued to endure and transmit itself.

Even with quarantine and public health measures, the flu took on a life of its own. Most insidious of all, the virus presented itself such that its early stages could easily fall under the banner of less serious, yearly, conventional flu. The sneeze or cough was at first little more than an inconvenience, a subtle but incessant probe through one's chest. Nevertheless, the world war continued on while the microbe flourished, far more deadly than any weapon employed by humankind. Although there is little reliable data on how many people were infected with the flu virus between 1918 and 1919, researchers estimate that approximately 500,000 people in the United States, and an estimated 40 million in the world, died from the pandemic. Influenza wasn't a new disease; instead, the conditions of the early twentieth century, which put a higher number of people in close proximity, increased the opportunities for the flu's circulation and unsentimental wrath.

The Life Cycle of Influenza

Early versions of influenza surfaced many times during human history. In some cases, medical philosophers classified the disease according to the plagues of their time, and in other circumstances, it became a metaphor for war. During his lifetime, the ancient Greek philosopher Hippocrates compiled a catalog of infections in his *Corpus Hippocraticum,* which provided insight into the etymology and progression of various diseases. Hippocrates writes of the "Cough of Perinthus," a fifth century B.C. epidemic that swamped a Greek seaside town on the Sea of Marmara, afflicting the upper respiratory tract in what is thought to be the first influenza epidemic in human history.

But flu-like symptoms had pre-modern roots and these diseases' social and linguistic imprint carried on throughout the early modern period. During the sixteenth century, Italian physicians identified a condition they called *influenza di freddo,* meaning "the influence of the cold," a condition they later described as a chilling effect on the body. While the record is not clear on whether this pre-modern definition mapped onto the medical description during the classical period, influenza terrorized humankind frequently throughout the subsequent century. The earliest account of an influenza epidemic in the modern era appeared in Britain in 1803. Dr. Richard Pearson, a London-based physician who conducted observations of the epidemic during the initial phase, affirmed: "The Influenza, as it appeared in 1803, is precisely the same disease which has extended itself at different periods for near a thousand years." By 1831, a lethal strain swept across Europe with successive waves in 1833, 1837, 1847, and 1889. The recurrence of these outbreaks suggests that influenza slipped through the ever-growing cracks of society—ongoing class war, mounting anti-colonial strife, and concurrent disease crises, such as cholera. Prior to the development of germ theory, the waves were perceived to be random acts of God, but with time, and in some cases, people began to ascribe disease to specific places and people.

As Patrick Berche recounts in his article "The Enigma of the 1889 Russian Flu Pandemic," the 1889–1892 flu outbreak was "the first pandemic of the industrial era for which statistics have been collected," an indication that governments wanted to account more thoroughly for emergent infection and mass death. When the 1889 outbreak initially struck the Russian Empire, even including Tsar Alexander among its victims, many aspects of Russian life came to a halt—with factories, ports, and schools temporarily closed. Eventually, the disease spread from Russia to Western Europe, arising in Berlin, Copenhagen, and Paris. In 1891, the northern English newspaper *Yorkshire Evening Post* reported: "In some cases there are three, four, and five sufferers in one house. Nearly all the members of the medical profession are actively engaged day and night in visiting, consulting and dispensing, and occasionally their surgeries are besieged by persons seeking advice and waiting for medicine for patients." As physicians worked in these hospitals, they faced many obstacles—patients who cried over abdominal pain or children confused by their lethargic state. Although it was not as destructive as the 1918–1919 flu outbreak, between 1889 and 1894, the flu killed nearly 100,000 British people. The trouble with viruses before the twentieth century is that they had invasive tendencies, startlingly brutal in causing harm.

In the midst of massive uncertainty and mortality, imaginations run wild. One rumor blamed a novel technology—electric lights—for the contagion. At times, even early media was an outlet for false information. On January 31, 1890, the *New York Herald* ran an article suggesting that illumination from railway cars and steamship cabins was somehow responsible for a global influenza outbreak, noting, after all, "the disease has raged chiefly in towns where the electric light is in common use." The article also noted that the disease "attacked telegraph employees." Like many statements that lacked evidence, the tale played on people's fear that innovation, leaps apart from and beyond tradition, could be the source of bodily harm. Beyond that, the economic depression of the 1890s, juxtaposed with growing discontent with authority, cast doubt on the general state of society at the end of the nineteenth century. The world stood in shock as a disconcerting wave of war and disease swept across even the healthiest people in European society.

At the outset of World War One, influenza had a different life. Thousands of soldiers who served in the trenches in the Rhineland or resided in barracks in Eastern Europe contracted influenza and were hospitalized. Sore throats, headaches, and fevers joined in as fighters suffered from ghastly wounds incurred during the battle. Unlike bullets, the flu did not just attack flesh; it unsettled the mind because of its overwhelming power over the body; the cold sweats rippled on the skin's surface as strength quietly receded below. Doctors monitored, treated, and theorized about the life cycle of influenza while also contending with a government that wanted nothing more than to deflect blame for disease run amok.

In societies where fear could be spread through text, governments counteracted by shaping the flu narrative. Under the 1914

Defence of the Realm Act, British media were prohibited from printing or spreading information that might "cause disaffection or alarm." On the surface, the legislation was sensible by trying to minimize collective stress. But in reality, it unlatched the doors for censorship, particularly about the origins of the outbreak. So long as Britain was not considered the origin of the flu, publications were permitted to write as they wished. During the early months of the 1918 influenza outbreak, the British media—similar to other Western media outlets—referred to the disease as the "Spanish flu," a denotation that bears resemblance to the 1889 "Russian Flu" outbreak.

This work to otherize the malady—as a foreign invasion—neither reduced anxiety nor prevented the illness; instead, it perpetuated a message that the flu was an uninvited guest overstaying its welcome (and perhaps of less concern to England's domestic population). Nevertheless, in Britain, similar to the United States, soldiers disproportionately fell ill to influenza, which meant that the population associated the contagion with militarization. A 1919 piece in the *Manchester Evening News* reveals an unfortunate and mislaid hope that human bloodshed and conquest could defeat the virus.

The end of the war did not end the Spanish flu. As the death rate soared, the joyful crowds gathered to welcome the Armistice in Albert Square, Manchester, unwittingly inviting the Spanish Lady to join them. The killer virus remained active well into 1919.

Even when the war ceased, influenza's reign did not wane. The global impact and power of 1918 was unrelenting and exposed a fester in the public health practices in even the world's mightiest and wealthiest nation-state of the time. In 1918, well before the establishment of the National Health Service, Britain proved ill-equipped to maintain its people's health in a time of catastrophe. The outbreak caused the government to shift its policy from one of relative inactivity to a suite of diverse tactics, led by a patchwork of charity organizations and government initiatives.

British doctors implored radio programs and film productions to warn people about the dangers of the influenza. One of the most enticing British films campaigning for public health was an eighteen-minute silent production, *Dr. Wise on Influenza*. The video begins with a grainy shot of an elderly man standing in front of a microscope. Like many broadside films of the time, the moving pictures were interlaid with a message. Throughout the program, we see people sick with the flu, sneezing on others, taking public transportation, and unable to get out of bed. The flu takes a toll on their bodies, but as we see, the virus also spreads throughout the city. But unlike an apocalyptic film where the viewer develops affinity for the character, the cinematic program provides directives. In one card, Dr. Wise informs the audience that "All infectious diseases, such as influenza, are caused by the invasion of microbes," and in another he recommends that a flu patient treat himself by, "gargling his throat and douching [sic] the nose with potassium permanganate and salt." Public health messages such as these were modestly funded by the Local Government Board in Britain, offering a visual aid on how to behave.

Regardless, societies would need more will and better coordination to tackle the flu. By 1919, the city of London centralized its public health system, partially through the establishment of the Ministry of Health, by expanding regular street cleaning and waste removal services. Initially concerned with child and maternal health, small as they may seem, these measures played a crucial role in growing the arsenal of government-funded initiatives to secure health and confidence for everyone. By the end of the decade, officials were reluctant to introduce quarantine restrictions on buses and trams for fear of damaging morale. Still, they took some time to conduct this collection for influenza.

We take for granted today the precision—perceived or otherwise—granted by modern recordkeeping and statistics for most of the West. We can record a birth through a certificate, or confirm a cause of death through an autopsy. But these mechanisms did not always hold in early 1919—especially for the working poor.

In November 1918, the *Times* reported a salacious article, "Triple murder and suicide," suggesting that murder and suicide were linked to the influenza outbreak. Aiming to enliven the dull rhythm of life, the paper veered slightly from the truth. That fall, Leonard Sitch, an avid baker in Suffolk, England, had a breakdown, stabbed his wife and two children, and eventually hanged himself. Some of his neighbors were in disbelief, citing that he was respectable, while others told reporters that his mental health was attributed to the aftershock of having the flu. Although some people were skeptical about the relationship between influenza and psychosis, at the time, some scientists believed there might be an association.

During the first part of the 1918 pandemic, doctors occasionally cited psychosis when reporting on the pandemic. In 1919, the British physician Dr. George Henry Savage believed that influenza deteriorated the nervous system, which could subsequently "originate any form of insanity." What he argued was that when people were bedridden, dark-eyed from the flu, time ceased. Savage was not alone in his thinking. Karl A. Menninger, an American physician, conducted a study in 1919 at the Boston Psychiatric Hospital and surmised a link between influenza and mental health disorders such as dementia and delirium. At a moment when physicians were trying to figure out why and how people were getting sick, they wanted to find a bridge between the predilections of the body and mind.

Mental illness—as lived by the people who suffer from anxiety and depression—is so distressing precisely because the person can become, as author Rachel Aviv notes, a "stranger to themselves." Derangement, depression, and suicide did coincide, the data says, with influenza outbreaks, but they were most likely a consequence of the social unease and material strain that existed in the early twentieth century. The lingering ghosts of world war, profound class division, and unaddressed traumas surely reinforced alienation and more grave psychiatric issues.

Disease response was fragmented and complicated. In 1921, the *Times* noted that London was still haunted by the pandemic, to the point that "our minds [were] surfeited with the horrors of war," which, in turn, resulted in even more carnage because of the "catastrophe" of the flu. Society was under immense pressure not only to attend to the dead but to find small ways to secure some agency for the living. On November 6, 1918, in southeast London, one superintendent requested that the local government provide him with twelve gravediggers because he had a backlog of corpses to bury.

Not every public health measure was morbid. Some found solace in even the smallest measures of control over a spiraling environment; residents in London's Hackney neighborhood, for example, were advised not only to isolate themselves in their beds if they had flu symptoms but also to rinse their mouths with salt and potato. In addition, some of the most controversial public health tools were enforced quarantine and surveillance. Public health methods were rooted in individual responsibility, even if they included practices that today would seem far-flung from germ theory; isolating oneself in bed, and engaging in homeopathy, was part of staving off disease. For industrial workers, whose wage was dependent on showing up to the factory, this was not always possible.

Contrary to popular thought, the 1918 flu would claim more lives than the bubonic plague. In Britain alone, that death toll was over 200,000. The inescapability of infection meant that people sought care wherever they could find it. The bed featured as a place where flu patients—whether in a public military camp or their private homes—found refuge. People's experience with the flu did not always look the same. For soldiers on active battle duty, most of whom were living in close quarters, social distancing was nearly impossible. But a section of the creative workers, many of whom came from the upper class, had more autonomy on how their home situation was structured. The flu was not only a phenomenon of mass contagion, which rumbled through entire communities; it was also a private battle that people felt in their beds.

The Creative Class

The flu crushed a stratum of the industrial class—soldiers and factory employees—and yet, another type of labor—that of a writer—could prevail. We see that those who could isolate, even if they were re-infected with the flu, did not only continue their craft, but were able to use the disease for literary inspiration. For the cultural elite, the 1918–1919 influenza epidemic wasn't merely a nuisance that ravaged their bodies and minds, it could be a source of creative motivation, and given their class background, loss of work did not interfere with their ability to survive. One of the best examples of writers who struck the balance between peril and inspiration was the British writer Virginia Woolf.

During the 1918–1919 flu pandemic, Woolf was dimmed by her body's decline and moved closer to a ghastly mental state, which affected her writing and relationships. Woolf navigated through cycles of hallucinations that were colored by hearing voices. Her physical precarity was blunted by her mercurial mental state. Woolf was perennially in bed, with mental and physical illness bristling at her side, while she agonized over her discomfort.

Like the soldiers at Camp Funston, Woolf's bed was where she was expected to find refuge, yet, given privileged class, her berth was private. The "sickbed" became a kind of liminal space for Woolf, both curtailing her ingenuity and also acting as a lodestar for that creativity. Nineteen eighteen was a turbulent year for Virginia Woolf's health—not only because of the flu but how her physical and mental health was slowly deteriorating. For Woolf, part of how she dealt with the illness was to write about it as precisely as she could. In her letters to family and friends, she reflected on her pain, noting, "My hands shake no longer, but my mind vibrates uncomfortably, as it always does after an incursion of visitors; unexpected, and slightly unsympathetic." Even when she suffered, Woolf's beautiful prose pulls one to her world. This was an unfortunate year. Some of her most difficult letters emerged from her reclusive state, sick with the flu in bed. This space of confinement, the bricks, and mortar, stilled for a recluse, left a lasting impression on her. Nevertheless, writing became her way to overcome that isolation.

Woolf never failed to document her health and that of those around her, though when her neighbor was infected, her timbre was initially dispassionate. In July 1918, she recalled that: "Influenza, which rages all over the place, has come next door," remarking that her next-door neighbor was infected. Besides this, she expressed a brief remark about a short stint of drought, with the city lacking rainfall for several weeks. Woolf nonchalantly remarks in her diary that a neighbor died from the flu. While her neighbor's infection did not directly impact her life, several months later, in January 1919, another physical adversity tormented Woolf. That month, she suffered from tooth extraction, and she spent a fortnight in bed, where she was unable to commit fully to writing. Her spells with influenza and concurrent diminishing health—at times—hindered her capacity to engage in literary life.

For the novelist, a bed can be a means to rest and regenerate. In December 1919, Woolf was recovering from the flu, feeling inarticulate from the infection; she described the illness with abnegation, "I was attacked—8 days in bed, down today on the sofa, & away to Monks House tomorrow." From bed, she sat stern, close-mouthed, and pensive, and while her symptoms were moderate, they were prolonged and left her incapacitated. Woolf's narration is ripe with stoicism, informing us of the unassailable truths about isolating in a bed while sick. She tried to salvage a delicate recovery period through a mix of rest and hydration. Regardless, her encounter with illness cannot be distinguished from her writing or her desire to understand how confinement slows bedside spirit. That day, on December 28, Virginia Woolf felt elated, untethered by the grim weather; instead, she ran through her bedside reading—*Greville Memoirs*.

For most working-class women in the late nineteenth century and early twentieth century, writing was a privilege that existed outside their station. And more specifically, writing about one's health was even more rare for proletariat women. Well before Virginia Woolf acquired influenza and was bedridden, her mother, Julia Stephen, wrote a slim catalog manual on how to attend to the sick and care for the suffering. Published in 1883, Stephen's treatise was meant not just to hyperbolize the tiny details of the sickbed but to highlight how attending to a sick person was not a leisurely act that could be carried out on a whim, but part of curating a hygienic sickroom, as well as a comfortable bed, significantly shaping the power people had to overcome death. As Stephen saw it, the art of nursing was predicated on meticulousness in tending to the room and bed so that the patient could have optimal health.

For Woolf—writing and illness had a less circuitous path. As an adult, she struck a balance between her bohemian life in London and her semi-hermetic life in Richmond. Despite being married (or perhaps, mainly because she was married), Woolf indicated that isolation fueled her imagination more than the company of others. Through her association with the Bloomsbury group, a collective of English intellectuals who lived in central London, she found community while re-shaping English literature and engaging in brief trysts and ad hoc salons. While not a formal school, most members were cut from the same cloth, both in their class status—upper class—and the university they attended—Cambridge. As elite as these people were, their lives were dictated by their rebellious passion for a non-conformist existence. For example, Woolf was well known to engage in affairs with other women outside the confines of her marriage. In her diary, she repeatedly describes her relationship with these writers—their successes and flaws. In a cavalier remark about some members of the Bloomsbury group she noted: "The classics make the time pass much better than the *Paul Mall Gazette*. Maynard Keynes came to dinner. We gave him oysters. He is like quicksilver on a sloping board—a little inhuman, but very kindly as inhuman people are. We gossiped at full speed about Adrian & Karin (Adrian's lovemaking done in loud judicious tones) & of course Majorie & Jos." These short, sometimes vacuous remarks were crystalline: Woolf loved to chat about scandal, a pastime that filled her with glee. In spite of a free-spirited lifestyle and these occasional moments of lighthearted bliss, Woolf would also admit to her mental health issues; in some cases, she expressed her instabilities to her close friends.

Writing to her friend Violet Dickinson in 1904, Woolf said, "All the voices I used to hear telling me all kinds of wild things have gone." By Woolf's own admission, these moments of mental vulnerability posed a barrier to her writing. They interrupted her drumbeat rhythm for writing and editing. But, given this instability, she used her solitude to lean back into literature. Tethered to her discipline, composition brought her elation even if she did not believe her work was stellar: "I wrote all the morning with infinite pleasure, which is queer, because I know all the time that there is no reason to be pleased with what I write." Like her contemporaries, she sought to draft the limits of recovery during the waxing and waning periods of the flu.

When a third wave hit the United Kingdom in the early part of 1919, Woolf contracted influenza again. This time, letting her sister know that her body "won't let [her] get up." Unnaturally fragile, with a still disposition, flooded with a perennial flow of headaches, tooth extraction, and mental anguish. Unable to distinguish between the demise of her strength and the burgeoning mental anxiety, Woolf added: "I thought I was probably dying, but Fergusson says it is only the nerves of the heart go wrong after influenza." During this period of infection, her illness wasn't just a matter of course; it interfered with her craft. In her article "View from the Sickroom," the scholar Janine Utell suggests that Virginia Woolf's writing was shaped by illness, even in her capacity not to engage in the practice: "writing offers a structure to the self until illness overwhelms it." Her diary entries, at times fragmented and sometimes cogent, are an exquisite description of how creatives retrogress at their talent when their body fails them. Perhaps Woolf suffered from severe migraines and was not merely exhausted by the flu, but by the procession of extraneous voices and her melancholy.

Erudite in both form and thought, Woolf was remarkable in attention to detail; she conjured language by channeling plain intro-spection, revising her sentences through economical prose, "here I am chained to my rock: forced to do nothing . . . every muscle tired, and the brain laid up in sweet lavender." From her bed, Woolf nar-rates with a measured interiority, revealing her fragile mental state, which is both raw and inert. What looms large is her deconstruction of herself and how literature became a way for her to stave off her depression.

Writing brings out one's insecurities, even if there are plenty of examples of accolades. In October of 1920, Woolf observed,

> Melancholy diminishes as I write. Why then don't I write it down oftener? Well, one's vanity forbids. I want to appear a success even to myself. Yet I don't get to the bottom of it. It's having no children, lives away from friends, failing to write well, spending too much on food, and growing old. I over-think of whys and wherefores; too much of myself.

In an era when living has become expensive, these anxieties ring true for many writers today, who navigate the precarity of compos-ing when the social pressure to produce and the material stress to survive come to a head. Although Woolf was successful in her life-time and not lacking in funds, the more she got ill—both physically and mentally—the more she doubted herself. Needless to say, her anxiety was also tied to her constant struggle to recuperate.

I could not fathom how Woolf's experience with the flu contin-ued to pierce through her social life, often leaving her bedridden and away from her art for many days. In January 1922, Woolf con-tracted influenza again; this time, she "was shivering over the fire & had to tumble into bed with the influenza." Woolf was sick for one month and could not work. There was something about that year, where one affliction followed another one, which left her unnerved, and constantly going back to bed.

Biology alone does not shape the experience of being indisposed; being secluded might disrupt the process but one can use the craft to avoid another task. Between April and June of 1922, Woolf admits that "not a word has been recorded. And I only write now to excuse myself from copying out a page or two Jacob for Miss Green." Being sick disrupted her writing flow, but losing her teeth caused her distress to the point that she believed it "may be some sort of cause for my ups & downs." For weeks, she was devastated, but over time, she found connections between the flu and war.

Writing in the aftermath of the First World War, Woolf saw the threat that the flu of 1918 posed to the stories of national triumph. Still, once she recovered from this period of social isolation by the summer of 1922, she drafted "Mrs. Dalloway in Bond Street," the precursor to the novel. Perhaps writing a novel was not a way to expunge the quiet suffering, but instead, to sketch the fidgeted unease and the calculated relationships of this period. Literature offered her a chance to bestow narrative beyond her body, psyche, and sickbed.

Illness as Metaphor

During the 1918–1919 flu outbreak, trauma was the bond that war and flu survivors both shared. Without aiming for a totalizing truth, Woolf pulled toward literature to convey this leitmotif. In her seminal essay "On Being Ill" she wrote that "those great wars which [the body] wages by itself . . . against the assault of fever" still linger in the hearts and minds of the population. The text is a stellar specimen of cultural criticism, ripe with a philosophical intervention about illness narratives. Initially written for T. S. Eliot's journal *Criterion,* "On Being Ill" was later published with Hogarth Press, which she co-founded with her husband, Leonard Woolf. Her exposition was meant not only to describe her bodily transgressions during her illness but also to be a sharp-eyed philosophical guide for bibliophiles. She originally composed the extended essay from bed, in 1926, after one of her mental breakdowns. But the bed is not the central factor; instead, she made a conscious provocation, "illness has not taken its place with love, battle, and jealousy among the prime themes of literature." The text is a meditation, but Woolf worries, waiting anxiously for literature to take heed. As such, she believes the intricacies of maladies, such as bodily and mental transformation, need more attention from her fellow literati. So, she provided a vivid picture of the sickroom and of how contagion, whether acknowledged or not, shows the body's vulnerability to a microbe.

"The experience," Woolf writes, "cannot be imparted and, as is always the way with these dumb things, [the ill person's] own suffering serves but to wake memories in his friends' minds of their influenzas, their aches, and pains which went unwept last February, and now cry out, desperately, clamorously, for the divine relief of sympathy." She plods along, gnawing at one's essence, not with shock and awe at the constant failures of literature, but an offering to provide some imaginary contribution. The text's laser focus speaks to the early twentieth-century perspective on illness and echoes how we sit with our present moment and find some reprieve for ourselves, even when our vital force is at its lowest. Practicing her own theory, Woolf decided to incorporate illness into her own tale.

In the spring of 1925, Virginia Woolf published *Mrs. Dalloway,* a multi-threaded story that converts pandemic trauma into personal drama. Set in the 1920s, the text is mainly preoccupied with a day in the life of an upper-class British woman with immaculately crafted interiority. Woolf conjures a dinner party among the London literary elite: cutting into people's socialization, what they are eating and drinking. Characters live in a punishing climate, where they are reckoning with the aftermath of illness and war. Like some people of their time, they are flawed nobles. Clarissa Dalloway, the novel's main character, has a shifting personality, a demanding presence, arrogance, and prudish nature, which ultimately keeps the reader hooked on the tensions between herself and her growing psychosis. Her horror and detachment create a space for empathy, but more than anything, they slowly reveal how the flu wrecked her.

Although *Mrs. Dalloway* is candid in its spare dialogue, with descriptive language, it makes clear how influenza and the First World War affected the body and the mind of its protagonists. The novel illuminates the flu's indelible mark with acuity through the narrator's voice: "Clarissa was positive, a particular hush, or solemnity; an indescribable pause; a suspense (but that might be her heart, affected, they said, by influenza)." But it didn't stop with the heart. The aftermath of Clarissa's state mattered gravely insofar that "since her illness, she had turned almost white." With tender and lovely prose, Woolf manages to bestow Mrs. Dalloway's vulnerability, an attempt to consider a person's demise even when they are abundant in health.

The novel mainly focuses on Clarissa Dalloway, to show how fragmented survival can be. Woolf takes the reader through a visual ride of the city, introducing us to unsettled and solemn characters; these people are questioning their relationships and the limitations of their bodies. The fifty-two-year-old flu survivor, Clarissa, is remarking on her body and refers to it as "cool as a vault." As the reader slides through the text, her ventures between hallucination and reality cannot be fully cemented in her world. Later, when she walks upstairs in her house, her body feels "suddenly shriveled, aged, breastless." Blurring her physical state when she contracted the flu in 1918 and years later at the party, the novel establishes illness's mental imprint on people's lives.

When I re-read *Mrs. Dalloway*, I was gripped with deep pleasure and found myself bound to the page, especially to Mrs. Dalloway's character from her vantage point and through various characters belonging to the English upper-class world, but it is more than that. These characters were shell-shocked and had to reckon with the physical residues of war and flu and the constant trepidation of injury and death; Clarissa was apprehensive that a virus would re-emerge and bring her to an end. The text suggests that seclusion are neither private acts of bravado nor do they trigger melancholia in the masses. Instead, the book explores what it means to be re-triggered by influenza's prospect.

The novel has the ability for a character to stand in for millions of people who have gotten sick, their shriveled bodies, their pain, and even their death. We see how a post-pandemic life never allows certain people to feel reprieve. And yet, one of the most striking ways that Woolf treats the zones where the sick are held captive is through Clarissa's former sickroom: "There was an emptiness about the heart of life; an attic." Rather than pedantically crediting psychosis solely to the flu, she slips into Clarissa's stream of consciousness—albeit through a narrator—with sly ingenuity. Woolf goes on to describe the shape of the bed, Mrs. Dalloway's activity on the bed—reading; and how she slept—poorly. The lingering torment of post-pandemic life might seem like an unpleasant reflection on a mid-summer day. But the attic was a reminder of her illness precisely because this was where Clarissa felt sick to death.

This behavior shows that her illness was not just something of the forlorn past but a state that continued to be communicated through her body and a private room in her home, signaling the ongoing wound she lived through. But more than anything, Woolf unveils how her characters grieve the people they lost and the people they were. Clarissa and her peers are people who, on the surface, maintain their insular circle. Despite their perseverance, they cannot escape the cataclysmic aftermath of the virus. Her novel is ponderous and precise, tending to create some distance between where people were hanging on to the lingering aftermath of the flu, where they had been, and where they were afraid of heading. Woolf bends our understanding of how contagion can embody every aspect of our being. Laced with flowery language, the novel documents the space between her past, the constant infliction of influenza on her cheeks and heart, and the unprecedented trauma of living through a pandemic, but it does something more.

Critics have noted that *Mrs. Dalloway* is about trauma and recovery, the art of mourning when a society has been stricken by war and infection. Encumbered by their prickly state, the lingering anguish is not just about the nebulous understanding of a virus but part of a broader malaise of the twentieth-century fears. In *Mrs. Dalloway*, Woolf narrated the aftermath of a global viral pandemic by distilling the unbearable residues of anxiety with quiet lightness.

Like Woolf, her contemporaries offered their literary take on the flu pandemic. Katherine Anne Porter's 1939 novel, *Pale Horse, Pale Rider,* provided a swelling trilogy that traced a young woman's memory of the 1918 influenza pandemic, which laid out, in plain language, what it meant to survive. When the unnamed protagonist wrestles with the objects in her domestic space, she confronts the intimacy of death: "Too many have died in this bed already, there are far too many ancestral bones." Given the ubiquity of death, other flu-inspired novels narrated the tragic pain of loss. Published in 1929, Thomas Wolfe's *Look Homeward, Angel* functions like auto-fiction, a coming-of-age story that depicts the grief sparked by premature and flu-induced death. "Fiction is not fact," he writes, "but fiction is a fact selected and understood; fiction is fact arranged and charged with purpose." These narratives speak to the terrible loss that pandemics had on society's upper and lower echelons. Even in panic, literature offers twists and turns while providing vivid imagery.

By the mid-1920s, most Europeans learned to live with influenza. Haunted by the aftermath of the pandemic, some people had recovered; others had not—the connection between the flu and pandemic literature. Influenza became a twentieth-century pandemic, spread by people's breath drifting from one body to the next. Like many people who encountered Virginia Woolf, the cultural critic Merve Emre is right to point out Woolf's literary accolades. In her annotated version of *Mrs. Dalloway*, Emre articulated the significance of Woolf's writing characters and the possibility they offered: "We see it in the characters themselves, their minds made to kindle and glow before us." The reader not only connects with them for the pandemic they survived but how that survival categorized who they became. Woolf understood that it wasn't just the case that one constructs a pandemic plot but that one offers a mirror to society and a screenshot of our souls. That feeling is not equally felt.

The African American novelist James Baldwin once remarked: "All art is a kind of confession, more or less oblique. All artists if they are to survive, are forced, at last, to tell the whole story; to vomit the anguish up." When writers ruminate about illness, they engage in a metaphysical inquiry about the body and soul. After her cancer diagnosis, the American philosopher Susan Sontag unearthed the social dimensions of cancer and tuberculosis in *Illness as Metaphor,* showing how one condition offered sympathy while the other was tied to failure. But what was more significant in the text was her intervention about how people's perceptions of a disease sculpt how they relate to each other. Ultimately, the overt and covert ways that our maladies shape the types of humans we are expected to be. Some diseases, Sontag argued, have more metaphorical weight precisely because of the fear that comes with them. An ailment like cancer "was never viewed other than a scourge; it was, metaphorically, the barbarian within." Sontag described, with sharp precision, how disease can have more to do with the psychological will of the person rather than the physiological severity of the disease itself.

Illness, especially in bed, constrains us, creating a permanent divide between the indisposed and the healthy. Decades apart, Woolf and Sontag elucidated that one cannot be shown what illness is like through remedy alone; you must work through it by writing. Reading contagion through Woolf is like ripping through a body-centered vision of the world and makes a case that we need to unpack the robust damage done when illness cuts through our core. But even more so, the text precisely shows how Woolf stammers through the powerlessness of isolation. Confinement, under these conditions of ill health, might cause damage to literary life, or reveal how pathogens leave us unmoored.

In 2022, writing for the *New York Times*, the journalist Alexandra Alter inveighed against the pandemic plot, arguing that "The relatively glacial production cycle for fiction is also creating obstacles for writers who, like the rest of us, can't predict what life will look like in the coming years, and worry that pandemic references might make their novels feel dated." For Alter, even if plagues have caused people to be prickled with fear, living in the time of plague has not always produced the most compelling literature. While elements of this thesis might be true, I disagree. Some seminal novels are worth revisiting mainly because they provide room for ingenu-

ity, strangeness, and insight into another world, showing how literature bridges society's flaws.

Writing about illness in literature can be a way back to recognizing our mortality, the awareness that our bodies will fail us. And
that realization—especially when depicted in fiction—transcends
the page and spans across generations. Virginia Woolf tried to put
forth how the early twentieth-century flu pandemic was a story of
survival, of those who prevailed and those who appeared restless.
Flu victims were navigating the world in a time when British society became more intertwined in an industrialized world, marbled
by a still-fluttering technologically driven death. But more than anything, Woolf's reflections on illness and the sickroom marvel at the
emotional luster of people whose worst fears do not emerge from a
vacuum, but people are deeply aware of their flesh and mind re-tell-
ing an encounter from the past.

More Than a Bed

Being sick in bed can lead one to reflect heavily on the body's limits.
If there is anything one has learned about contemporary women
essayists who have been diagnosed with cancer, it is that writers are
willing to utter their sentiments about death. Many, in their attempt
to reckon with the delicate balance of illness and sickness, use their
writing to unbend the afflictions on their physique and the contours
of their confinement. The very act of women writers reflecting on
cancer elicits a particular narrative about the difficulty of work and
the nature of rest. As the Black feminist Audre Lorde transcribed in
her journal six months after her cancer diagnosis, "I'm not feeling
very hopeful these days, about selfhood or anything else. I handle the
outward motions of each day while pain fills me like a pus pocket

and every touch threatens to breach the taut membrane that keeps it from flowing and poisoning my whole existence." At the time, she was fifty-eight years old; with each passing day, Lorde was convinced that her doctor's prognosis was a death sentence. For months, she imagined how she would die.

More recent accounts of cancer's terror cite the bed as a realm where thoughts about death are activated. In her cancer memoir, *The Undying*, Anne Boyer postulates, "A sick person in bed is the ward of love if she is lucky, and the orphan of action, even if she is not. All the accumulated gorgeousness of life in bed can be eclipsed by gravity there, and dreams, too, become occluded by pain. During illness, every pleasure of a bed can disappear behind new architectures of worry." Boyer asserted that the bed, where she sought sanctuary, was also where she contemplated the mode and meaning of her mortal exit. Boyer's articulation of the tragedy of the bed is relatable, and all three writers—Sontag, Lorde, and Boyer—were writing precisely during a moment of treatment, near death, and remission. The sickbed is not only a physical space where writers have a trope for understanding confinement; it also provides people with philosophical insight. Though all these author's were generations apart, Woolf's irregular warfare with illness and recovery also observes the emotional toll of confronting one's mortality.

While writers diagnosed with cancer incorporated accounts of feminist writers scribbling from the bed, recent compositions of bed writing have constituted a space for coronation for disabled and chronically ill people. As the Haitian American artist Carolyn Lazard succinctly put it, after spending two years lying in bed, the room also presented itself as a place of scientific inquiry, a way to heal the body from the disease. Lazard's essay reads like this: "From my hospital bed, I would look up alternative treatments and scroll through Crohn's forums on my smartphone." For the author, the allure of writing from bed when one is chronically ill is not merely utilitarian; it can give the artist the possibility to exercise one's agency.

This act of discovery and renewal is part of the reflective power of other queer creatives; for some, tone and narrative reinforce a sense of affinity. As a non-binary writer, Leah Lakshmi Piepzna-Samarasinha had trenchantly expressed the significance of writing from bed: "When I moved to Oakland in 2007, I started writing from bed. I registered in old sleep pants, lying on a heating pad, during the hours I spent in my big sick-and-disabled femme-of-color bed cave. I wasn't alone in this. I did so alongside many other sick and disabled writers making culture." Piepzna-Samarasinha offers solace, a zealous charm that tends to others who have labored from their divan. The bed is not just a space of confinement; it is connected to a lineage of disabled creatives—such as Frida Kahlo and Grace Lee Boggs—who were engaged in artistic and social movements in spite of their impaired states. To lie in bed can be a way to turn away from one's physical disability and test one's capacity to create and transform one's art practice. Bold and jarring, these people provide prescient and frank interventions about the sickbed, not solely as a site of physical constraint but as a place of scientific retrieval and shimmering hope. But this assumes that everyone has a residence to begin with.

There is an unevenness in who has access to a home, whether healthy or sick. And the conditions that force one to be in bed for an extended period, such as a prison or a mental health clinic, might be mired with unbearable pain for some of these institutions' lack of care. For the incarcerated, the relentless violence of the prison, trapped by the slow tick of passing time, might mean that the mattress is the only site of comfort. Nevertheless, this dwelling place might be a sanctuary that provides an environment to challenge one's inner demons or the seismic waves of grief. Having a domicile during an epidemic can make a difference between survival and death. As with cancer, people who had access to a bed during the 1918–1919 flu epidemic illustrated how important rest was not only for one's health but also in diminishing the spread of the disease. The history of the flu's destruction has fueled the innovations entailed by researchers.

It was in the 1930s that scientists began to advance their efforts to tackle the flu, which suggested that humans could and might have a future without the disease. The quest to find a vaccine was developed with Thomas Francis and Jonas Salk in 1942. Since then, the flu vaccine has mutated, and many humans have adapted to its presence. Even if the magnitude of the 1918 death toll is long gone, scientists are investigating its molecular qualities. By 2001, researchers at Mount Sinai Hospital in New York City had sequenced the influenza genome, finding links between classical iterations of the disease and the early twentieth-century pandemic. Their work— and the labor of other scientists—suggests that a particular viral gene, NS1, possibly mediated the virulence of the modern version of influenza. In addition to being a heroic act, their efforts to use science are not in vain, mainly because the flu continues to be a source of discomfort for millions of people worldwide.

According to the World Health Organization, the flu kills up to 650,000 every year, often harming the most vulnerable—the young, the elderly, and the immunocompromised. On the whole, the world has learned to live with the flu in ways that Virginia Woolf's contemporaries could never have imagined. Given how this number pales compared to a century before, it is easy to see how even the most moderate manifestation of the flu can be seen as innocuous. We assume that reinfection is asymptomatic, but for Woolf, it was not. As such, the impact of the disease is not just about how we understand it today, but people's experience and memory.

Woolf is renowned for her cogent prose, which delivers a scene to the reader's imagination. She transferred that energy in her diary, writing ferociously from her pen. Unbelievably sallow in complexion during her periods of malady, Woolf also remains biting on the page, often documenting her torment with mental illness. But her seclusion and bedridden state meant that she read profusely. That confinement was a period of literary transference.

During her lifetime, Woolf was well regarded for her literary genius, an essential element of the early twentieth-century cultural milieu. She commanded attention because she excoriated the customs that kept so many women bound to the domestic labor of marital life—cleaning and childcare. Later in her life, Virginia Woolf continued to be tortured by her mental illness; by her account, the voices were too much for her. Her refusal to be alive meant that she would determine her course. Her suicide note, which she addressed to her husband, Leonard, revealed her torment: "I can't fight any longer. I know that I am spoiling your life, that without me you could work." On March 28, 1941, shortly after composing this letter, she exited her Sussex home on a crisp spring afternoon and walked to the River Ouse. Perhaps she passed by the deciduous flowers in the fields, blossoming with comeliness, or the mossy trees near the riverbank. Maybe she doubted her actions briefly and stood motionless with brooding stillness, a serenity that comes when a person has decided to become their executioner. That spring day, at the age of fifty-nine, she drowned herself in the river.

In concert with her work, the bed was also a place for free association, a zone with a strong sense of interrogation. Writing portions of this chapter from my own bed was my attempt to create a new sense of self. Typing upright, with my computer on my lap; legs snug under my blanket, and a construction crane carrying concrete blocks outside my window; every other morning, it is my partner's turn to make our coffee, during which time I began admiring the possibility of connecting with the bed, not just as a place of confinement or my tortured battle with insomnia, but the potential for more—for intellectual inspiration. One gets a sense in Woolf's last moments, far away from her home and bed, that death was the only way she would no longer be held captive by her illness—dying would set her free.

The bed is both a metaphor and not.

Chapter 4

BREAKING THE WALLS OF SILENCE

KATHY BOUDIN SPENT MOST OF the first twenty years of her son's life in prison. She was denied the privilege of seeing him learn his first sentences or ride his bicycle; instead, she was an inmate prone to the merciless attention of New York State authorities. She was a slight woman, poised and often deep in thought, a seasoned anarchist and avid poet. Her partner, David Gilbert, expressed that Boudin was "an incredibly compassionate, talented, and thoughtful woman," praising her acumen inside and outside the prison. During the last decades of her life, she was a formidable critic of the prison, stressing the grim conditions in which she and her fellow incarcerated inmates lived. Iconic for her political radicalism and beloved by her comrades, Boudin stood apart from most people because she always leaned toward a collective responsibility for social problems.

Born in New York City at the height of the Second World War, Boudin was outfitted with a lively Marxist education that blossomed for decades. The child of the notable progressive lawyer, Leonard Boudin, she was exposed to left-wing politics early on in her life. Her heroes included such people as African American thespian Paul Robeson and Cuban revolutionary Fidel Castro. What she admired from these activists was that they embodied and affirmed an internationalist approach to social movement. By the time she entered Bedford Hills Correctional Facility in New York State, she had already established herself as a vigorous advocate for the working class and poor. As her sentence drew on, she became a devoted ally to her HIV-positive comrades in prison. That role shifted as she went from ally on the outside to fellow prisoner.

Like many radicals of the 1960s, Boudin saw organized social movements as righteous *affrontement* against the exploitable class. Initially inclined to the liberal organization Students for a Democratic Society, she eventually co-founded the Weather Underground, a far-left Marxist group, in 1969. Like many radical associations during the Civil Rights movement, they were militants, both in rhetoric and tactic; the syndicate expressed interest in creating a revolutionary party to overthrow the US government. At the same time, their methods were deemed subversive by centrists and pacifists in America.

In the film *The Weather Underground,* documentary filmmakers Sam Green and Bill Siegel depicted the organization not just as a breeding ground for domestic terrorists—which is how the US government classified them—but also as committed freedom fighters who aimed to build a working-class movement. For activists of this generation, direct action was part of the modus operandi for armed insurrection. Despite doubts about the methods, then and now, about their political strategies, activists such as Boudin wanted an equitable world—where no one would ever be exploited. A politics that put the salt of the earth at the center.

Boudin's membership in the Weather Underground dissolved after she and her comrades were charged for their alleged participation in a failed heist. On a mild day in October 1981, Boudin and an accomplice attempted to rob an armored truck in Nanuet, a small village on the Hudson River in upstate New York. Her accomplice shot and killed two police officers and a security officer in the brawl that followed. Although she did not pull the trigger, she was vilified for her involvement. "Kathy Boudin is not a sympathetic figure," recounts the *New York Times*. "Although she has never been charged with personally committing violent acts, she was associated with people who have been convicted for attacking a Brink's truck in Rockland County and killing a Brink's guard and two police officers in the process." Well before the trial, the public already articulated their sentence. She was guilty for being a revolutionary. Given that, Boudin fervently embraced the latter part of the organization's moniker for a decade, by remaining underground and avoiding the state's retribution for the Greenwich Village townhouse explosion or other bombings believed to have been part of their doing. Eventually, she turned herself in and stood trial.

When Boudin was sentenced on May 4, 1984, the District Attorney prosecuting Boudin's case admitted as much, stating that she "did not fire a single shot and was unarmed," and did not participate in the planning of the robbery. Nonetheless, the punishment was severe, which was common for radical activists during the 1970s and 1980s. "I know that anything I say now will sound hollow," Boudin remarked, her sentence of twenty years to life confirmed, "but I extend to you my deepest sympathy. I feel real pain." Though Boudin was constrained by her incarceration, she worked toward something larger, believing that even in captivity, HIV positive people had a right to quality medical treatment. But more than anything, establishing a collectively run advocacy group within the prison that championed women's health was one of the many contributions she made during her incarceration. Boudin's radical vision— inspired by the pedagogy of Paulo Freire and James Baldwin—was predicated on imagining life beyond the walls of her confinement. "Many of us are looking for alternatives to the actions that brought us here. We are working to imagine new choices, to widen options, and to figure out how to make these real."

Boudin's Play Against HIV/AIDS

Bedford Hills is a typical northeastern American town. With lush rolling hills, colonial estates, and garish strip malls, the Westchester hamlet is home to three thousand people. Midway between the Hudson River and the Connecticut border, the settlement is a menagerie of dense woodlands sprinkled with oaks and cow pastures. Some would assume that Bedford Hills, by virtue of being along the Metro North train line, is an extended suburb of New York City. Near the edge of the municipality, Bedford Hills Correctional Facility is the largest penal institution for women in the state of New York. In the 1980s, like many women's detention centers, 20 percent of the women at Bedford Hills were HIV positive. Incarcerated women were one of the most vulnerable risk groups for the virus at the time.

During the first decade of the HIV epidemic, the virus's blight at Bedford Hills prison was far grimmer than in civil society. Between 1988 and 1994, AIDS-related deaths were the leading cause of death for African American women in New York State, which impacted incarcerated and formerly incarcerated members of this group. Bedford Hills was not unusual. By 1989, New York State had more AIDS cases than any other state. Like the women at other prisons in New York State, the women at Bedford Hills were disproportionately HIV positive. While the disease was associated with the gay community during the 1980s, incarcerated women suffered. Some of them believed they acquired the virus through intravenous drug use, while others believed the transmission occurred from an HIV-infected partner—either a husband or a john. What was clear was that some of the inmates were segregated by prison guards, without clarity about when they would be re-integrated with other inmates. Others wanted more information about the life cycle of the HIV virus. In the late 1980s, at the height of the epidemic at Bedford Hills, several of the prisoners took action to get clarity about the ailment and the impact it would have on the women in their facility. At the height of the epidemic, the AIDS Committee for Education (ACE), an organization at Bedford Hills that educated and counseled women who were living with HIV, was founded by three ardent activists: an HIV positive person and two former Weather Underground comrades—Katrina Haslip, Judith Clark, and Kathy Boudin.

When she first found out that she was HIV positive in 1987, Katrina Haslip was afraid to tell her fellow Bedford Hills inmates. That year, she was more than halfway through her five-year sentence for pickpocketing. Like many of the African American women at Bedford Hills, she felt that her prison term did not match the crime, but her conviction paled when compared to her HIV status. The prognosis was a death sentence. A former sex worker, intravenous drug user, and recipient of a blood transfusion—three risk groups benchmarked by the Centers for Disease Control and Prevention—Haslip worried that Bedford medical authorities had not tested her previously. Early on during the AIDS epidemic, HIV positive women at Bedford Hills, like herself, were initially segregated from other inmates, even though people knew that HIV could only be spread through blood and sexual fluids. Another issue that upset her was that she and other HIV positive women prisoners believed that they were being denied treatment. In the beginning, she admitted that she knew very little about the disease, but for Haslip, she was motivated to be an AIDS educator because she felt that "Women were dying in their cells and in the hospitals. They were dying because they were giving up and because they felt there was no hope." Co-founding ACE was a way to challenge that disenchantment.

In the early 1990s, Judith Clark described herself as a "red diaper" baby, an epithet for children of Communist activists. Growing up in Brooklyn in a Jewish intellectual household, she was ingrained early on in an anti-racist fray. In 1963, when she was a junior in high school, African American parents in New York City demanded that schools be integrated, asking white teachers and students to join their boycott. Clark and several of her peers joined the boycott, the genesis of her involvement in the US Civil Rights movement. Like Boudin, her fervor grew from the moderate activities in the Students for Democratic Society into more zealous actions with the Weather Underground, which included the Brink's truck robbery. By the time she went to trial, she defended herself and exclaimed to the jury, "Revolutionary violence is necessary, and it is a liberating force." Judge David Ritter, believing that she was unfit for society, sentenced Clark to a minimum of seventy-five years in prison. Between 1985 and 1987, she was placed in solitary confinement, after prison guards discovered her plans to escape Bedford. Bereft of her essence when she was in severe captivity, she slowly assimilated into a communal life through HIV advocacy. When describing her motivation for co-founding ACE, Clark noted: "Sometimes when you're handed something on a platter, you don't understand what it took to get it. Perhaps that is part of what led me to realize, when AIDS became a crisis, that we had to do something ourselves." Like the two other founders of ACE, Clark went back to her roots by assuaging the alienation HIV inmates felt from their family, friends, prison community, and society at large.

A duty to family was a central feature of why some Bedford Hills inmates joined the ACE Program. At its height, there were 800 people at Bedford Hills. Still, given how women disproportionately provide childcare—within and outside of prison—by the 1970s, the facility had to reckon with another factor: 85 percent of the 450 women at Bedford Hills had children. Many of them relied on an extended family to take care of their children while they were incarcerated. Others went through a metamorphosis, when they recognized some relatives were HIV positive.

For many members of ACE, their participation was a form of mutual aid, an opportunity to acquire information and camaraderie. Aida Rivera, a vital leader in ACE, saw her life transformed in ways she could not imagine prior. When Aida was arrested in 1983, she wasn't aware that she would be sentenced to fifteen years in prison. Raised in Brooklyn, she navigated a bilingual life, hearing her parents speak Spanish and responding to them in English. Her working-class Puerto Rican parents did everything they could do to sustain the family—taking jobs as they came, whether on an assembly line or in the junkyard. At sixteen, Aida dropped out of high school and took work at a factory.

As a mother of five children, her earnings were not enough to support her family, so she began a flexible occupation where she could make more money: She turned to selling drugs. Eventually, she was arrested for drug possession. Sentenced to fifteen years to life, Rivera knew she would miss her children's school plays or graduation ceremonies. The first three years in Bedford Hills were quiet, but when Rivera's sister tested positive for HIV in 1987, her subdued nights were replaced with work and study. As a result, she joined the ACE Program to learn about the virus and reconnect with her sibling. "ACE changed my life," she'd say. "I began to look at things differently. I started caring about other people, even people I did not know and might never see again in life." The sororal community at Bedford Hills provided the means to self-educate. After earning her GED and associate degree from prison, she eventually earned a bachelor's in education, but what mattered more at the time was the community and counsel Rivera could provide to those otherwise isolated and abandoned with what was, at the time, seen as a medical death sentence. Moreover, she had the ardor and determination of a Carthusian monk, and one of her first duties as a leader on the inside was to build relationships with HIV-afflicted inmates who could not advocate for themselves.

There is a chain of events that brought these women to Bedford Hills Correctional Facility, and what stitched their lives together was their eventual commitment to justice. People in the group were seeking different things. For some, HIV campaigning was the goal, and for others, it was prison reform. Similar in objective to their well-known and public-facing counterpart, ACT UP—the 1980s grassroots political group that sought to end the AIDS epidemic—ACE called prison authorities to account for the illness and death that thrummed through the queer and incarcerated communities. Unlike ACT UP, where members could stage massive street demonstrations and seek platforming from celebrity supporters, ACE predicated its work on patient empowerment and prison reform. Although both were difficult to achieve, they were able to provide group counseling for HIV inmates, they published newsletters to inform inmates about HIV, and they held educational workshops. But there were limits. The members of ACE found numerous faults with the prison system, which was taxing physically and psychologically, especially for people living with HIV. They ate, rested, and labored in the same facility, forming kinships that provided a balm against the new plague.

Although the exact figures were not fully known to ACE members, they gathered, implicitly, that the HIV-positive people at Bedford Hills were not getting the full medical attention they deserved, nor was there adequate emotional support to navigate through the fragmentary and elusive information about HIV/AIDS in the mid-1980s. Boudin encouraged us to have "a sense of efficacy and agency developed to all of us." The sheer compassion within the walls among inmates vastly contrasted with what New York State prisons exemplified during the 1980s. To even confirm a diagnosis was a hurdle. Access to HIV testing in prison was limited, meaning that those who'd wish to protect their fellow inmates, or themselves, lacked the basic resources to do so.

Education and writing were central features of ACE's work, a mild palliative in captivity. For example, Boudin described teaching a class at Bedford Hills where the incarcerated were asked to reflect on what they knew about HIV and whom they would talk to if they tested positive for the disease. Despite being an emotionally charged issue that brooded somewhere between perplexity and shame, there was perspicacity. "Instead of seeing AIDS as an individual problem," Boudin wrote, "people began to see it as a common one, and one they could work on together." Their publication, *Alert to AIDS,* was a marriage of their literary ambition and an acknowledgment of the emotional heights that fell when telling a lover that they were HIV positive, but more than anything, more than anything, it broke "the silence so that people could together begin to deal with the [AIDS] epidemic."

As a group, ACE did not just have the intuitive sense to address the HIV pandemic through their peer support program, they also challenged the racial palette of Bedford Hills and concluded that the US prison system was inherently racist. The New York State Correctional System, with fifty-six facilities, was the third largest in the United States. Of the forty thousand inmates, eighteen hundred were women, and nearly 70 percent were from New York City. Roughly half were Black, and slightly over one-third were Latina. The demographics themselves do not explain why these people were in prison. Half of the women at Bedford Hills were convicted of drug offenses, often because Black and Latino people were more likely to be searched and seized. Boudin, who at the time believed that she would spend the rest of her life at Bedford Hills, cared less about what brought people to prison and more about how to make them healthier and happier. Boudin sought to answer the question: How does one reclaim sovereignty over one's body when sequestered from the free world? For HIV positive people in prison, this was nearly impossible to conceive.

The African American artist Lorraine O'Grady once wrote, "What alternative is there really, in creating a world sensitive to difference, a world where margins can become centers?" Although intended to describe the subjectivity of African American women, this question also echoed what the women at Bedford were doing when they established the ACE Program. The question of who the nucleus of social justice is, both within and outside prison, forces one to think about the state of health in prison. Something happens when people see each other, not as individual subjects facing a chronic and life-threatening disease but as someone worth living and being documented—outside of a prison ID number. The history of the ACE Program at Bedford isn't just the story of Boudin, Haslip, Clark, or Rivera, but it is a chronicle of the history of the prison itself.

Prison Theory

Michel Foucault's *Discipline and Punish*, now five decades old, tells us "there is no glory in punishing," a message made even more clear in the light of today. The prisoner gets up when instructed by a guard and risks corporal punishment if they disobey the rigid timetable of the prison: strict mealtimes, regimented work, and unannounced searches. Decisions about how they use their time and where they can go are made for them. Perhaps Foucault's thinking can be extended further, to a sordid reality of modern life, an oblique state where our schedule is never ours even at a workplace or in a nuclear family. Nevertheless, the prison, Foucault posits, restricts one's life behind a barricade. It is a continually evolving entity, one not merely of laws and juridical practices that legislators establish; it is also a reflection of what and whom we, society, deem worthy. As he puts it, prison is intended to trouble the soul by glorifying prohibition and depriving privacy. Regardless of any injection of new regulation, Foucault believed—like many other radical scholars—that prisons do little to deter crime.

The field of philosophy is no stranger to the personal injury inflicted while in prison. Antonio Gramsci, a founding member of the Italian Communist Party, documented his theory of class struggle and cultural hegemony while incarcerated. His seminal oeuvre, *Prison Notebooks,* is a series of essays written during his own solitary confinement between 1929 and 1935 under the Italian fascist regime. The text, in part, details the deterioration of his body during his internment. And it also reveals what prisons are for, to exercise brutish and social isolation—both a political and psychological punishment tool. As his comrades piled up in Italian prison cells, he could not help but think that this space was meant to crush the spirit of counterculture. "I turn and turn in my cell like a fly that doesn't know where to die," he wrote of the vile hole deteriorating his sense and physique with trauma that will never fade.

Foucault's and Gramsci's perspectives about prison and confinement dissect and categorize the carceral state, unpacking confinement's role in shaping psychosis. But while prisons and solitary confinement provoke the sensation in the incarcerated, their captivity renders them on the edge of humanity.

Unlike these theorists, who offered insight about the state of prisons from within, ACE members focused largely on women's health and remedy, during the height of a viral epidemic. Boudin's reporting and the writing by ACE members were not just a surgical account that revealed the horrors of incarceration while HIV positive; they also demonstrated something far more acute. The imprisoned could offer each other a valuable message through collective care, rupture the walls of isolation in the most irreparable places. Reading through the stories of the members of ACE is an accounting, a way to see the jagged crevices in the carceral state, where friendships became palatable even in institutions meant to suck the marrow out of their bones. So many of the incarcerated women at Bedford Hills Correctional Facility were distrustful of the prison authorities, not in a facile way, but from knowing firsthand how institutions of all kinds had already failed them. For many of the one and a half million people incarcerated in US prisons, this was the first time many received essential medical procedures, such as a pap smear. A testament to the insidious state of the United States' lack of universal healthcare.

On November 9, 1973, while J. W. Gamble, a prisoner, was working at the textile mill at the Huntington Unit prison in Huntsville, Texas, a five-hundred-pound bale of cotton fell on him. He was distressed briefly immobilized, an acute pain gnawed at his back. Given the injury, he decided to go to the penitentiary's clinic, hoping to find some reprieve. Noticing that he had a herniated disc, medical staff attended to his needs. One physician, Dr. Astone, prescribed Gamble a pain reliever and muscle relaxant to alleviate his injury. As weeks wore on, his pain did not subside. In mid-December, weeks after his accident, prison staff expected Gamble to return to work even though he had excruciating torment. Rather than accept his claim, custodial staff retaliated and placed him in solitary confinement. During this period in captivity, his symptoms were aggravated, and by early February he suffered from cardiac arrest and was rushed to the hospital. Initially, Gamble was emotionally and physically deterred, jolted by his punishment for simply being unwell, but after his flirtation with death, he pursued a legal case. On February 11, 1974, Gamble sued W. J. Estelle, Jr., Director of the Texas Department of Corrections, for cruel and unusual punishment, citing the callous measures that worsened his condition. Although the state of Texas never reached a consensus, in 1976, the Supreme Court ruled in Gamble's favor, asserting that depriving incarcerated individuals of reasonable, adequate medical care violated the Eighth Amendment. The case established prisoners in the United States have the "right" to treatment, and treatment for the incarcerated became a centerpiece of rehabilitation efforts. Since then, prison reform advocates have highlighted that these amendments have improved the lives or health of incarcerated people.

Prison reforms, as mild as they were, would not have been possible if not for the organized resistance of the incarcerated. To understand how women prisoners work with each other is not a matter of juridical ruling or the extent to which it was exact; instead, it is to look at the politicization that organically happens to the detained. As Victoria Law describes in *Resistance Behind Bars,* women prisoners have collectively challenged custodial authorities—even if their efforts were not always widely known. Law highlights how incarcerated women at Bedford Hills—similar to the male prisoners who led an uprising at Attica in 1971—carried out the August Rebellion, a revolt where two hundred women protested the inhumane treatment of inmate Carol Crooks, in 1974. They revolted when the guards placed prisoners in solitary confinement without sufficient reason, arguing that prison staff used this measure arbitrarily. Despite this uprising having lasted just one day—the incarcerated women at Bedford Hills used their momentum to file a class action lawsuit, *Powell v. Ward*, arguing that segregating prisoners into a special unit without a formal complaint system contradicted the US Constitution's Fourteenth Amendment, which upholds that US citizens have a right to due process of the law. This meant that prison guards at Bedford Hills and beyond could be subject to accountability, and that incarcerated people had a right to challenge their solitary confinement.

Today, many activists would argue that reform is not enough. But a visit to Bedford Hills (or any US prison) years after the August Rebellion would show an institution still failing to provide adequate medical care to the inmates under its purview. In 1977, the prison administration was implicated by a judge for "the denial of necessary medical care for substantial periods of time." The early wave of HIV/AIDS cases showed that people who did not receive swift treatment were prone to experience the most rapid progression of HIV-related symptoms. This was a great dying, in full view of incarcerated and prison staff alike, a death rendered because of their confinement. As Kathy Boudin and her colleague, Judy Clark, noted in their 1990 study about Bedford, the prison—by its nature—could not address the AIDS crisis. The denial of fundamental liberties contradicted the premise of providing care. The crisis was glaring as 58 percent of the deaths of prisoners were from HIV-related illnesses. The immunosuppressed prisoners who were HIV positive also had to worry about tuberculosis and many other communicable diseases that went poorly treated and made the susceptible more likely to die.

As they witnessed these dynamics at Bedford Hills, prisoners became attuned to the tattered moral fabric that required a greater sense of care for each other. Since its founding in the late eighties, the ACE Program has not disappeared. The group continues to provide peer support, even showed heavily implicated in upholding the incarceration of HIV-positive people. But its inception and continuation raise a broader concern about women's health in New York State prisons and the dire need to endure even when one can barely survive. The HIV/AIDS activism at Bedford Hills during the 1980s and 1990s cannot be understood through the incarcerated alone. These were women who lived through the tail end of the American Civil Rights and Women's Liberation movements, and had witnessed firsthand how sexism and racism lived on.

Boudin was not the first to raise concerns about women's health in New York State prisons. Throughout the nineteenth century, the disposition of incarcerated women was a subject that caused anxiety, which sometimes coincided with prison reform. How women were housed gives us an entry point to the evolution of the prison as an institution and the substantial harm—both acknowledged and ignored—it inflicted on inmates' mental and physical health.

New York State of Mind

When Rachel Welch first arrived at Auburn State Prison, her punishment caused her world to shrink from the bustling city of New York to the asphyxiation of a bunker. For many European immigrants in the early nineteenth century, New York was a place to start over. The state had become increasingly relevant in domestic and international trade, already having easy access to Europe and now connected to the Midwest with trains to Buffalo and Lake Erie. That year, in 1825, Welch, a working-class Irish immigrant, became one of the first women incarcerated in a New York State jail. Auburn, one of the oldest prisons in New York State, initially housed both men and women who were charged and sentenced with a crime. William Brittin, the first warden of Auburn State Prison, prided himself over the "Auburn system," a penal method of the nineteenth century where prisoners would be subjected to solitary confinement or harsh physical punishment if they did not adhere to the prison's measures. In 1818, seven years before Rachel arrived, many of the inmates oscillated between strict labor in the prison's workshops during the day and strict silence at night. Dressed in grayish jumpsuits, the prisoners resided and provided cheap labor, making booms, buckets, and boots. Men worked in the ramshackle marble quarry. Women, on the other hand, were assigned textile work of picking wool and knitting.

But at the time, there was no singular place where women were housed after conviction or standard for how they were to be treated. Most of the women detainees were segregated and confined to an attic. In contrast to male prisoners who were placed in cellblocks of single cells, women shared a single room and slept in the same area where they attended to prison labor. Unlike most other women, Welch spent some of her time in solitary confinement, in a harsh, poorly lit cell. Although little is explicitly recorded about Welch, she was perceived to have violent tendencies. Dull day-to-day life in prison was generally intended to address the prisoner's alleged unfavorable characteristics, which for nineteenth-century women included rebellious spirit. But Welch's sentence—prolonged isolation—was meant to do something more. The extreme conditions were to raze a convict of her nature's nefarious core.

The ideology of solitary confinement echoed philosophers like Gustave de Beaumont and Alexis de Tocqueville, who observed, "Placed alone, given his crime, [the prisoner] learns to hate it; if his soul is not yet surfeited with crime and thus has lost all taste for anything better, it is in solitude, where remorse will come assail him." Despite this foundation of thought, there was not a similar presumption on whether women could be reformed by prison at all.

Here Beaumont and Tocqueville argued that "the reformation of girls, who have contracted bad morals, is a chimera which is useless to pursue." In an 1844 document, a New York State senator paid little attention to the women, believed that "female convicts were beyond the reach of reformation." What this meant is that scorn and paternalism were driving forces for how early nineteenth-century prison officials viewed women and their ability to be rehabilitated.

While early male philosophers and scholars had little to say about the effect prison had on women, feminist scholars have reflected, with sobriety, the sharper edge of captivity. In her articles "Prisons for Women," feminist professor Nicole Hahn Rafter argued "This special disdain for—even horror of—the female criminal helps explain why, in this period, the care provided for such women was usually inferior to that of their male counterparts." Similarly, Karlene Faith's remarkable book *Unruly Women* opines that prisons became a way to persecute women who did not adhere to societal gender norms. For most of the nineteenth century, Faith observes, the prison was where mentally ill or socially deviant women—not necessarily "criminals"—were housed, often against their will. The lack of consensus about whether prisons could repair and recondition women inmates for nineteenth-century society leaves one wondering why officials incarcerated these women in the first place.

Although nineteenth-century prison reformers believed that solitary confinement would be a path to penitence, there was little evidence that this system could induce remorse. A sudden, unexpected break from society may leave all sorts of issues between the prisoner and the community unresolved. Each prisoner experiences contrition differently, and very few records can provide, in full detail, the subtle and overt ways that incarcerated people in the early nineteenth century felt during and after incarceration. Instead, we know how they were neglected or, in some cases, abused during these periods of confinement.

In the spring of 1825, six months after her arrival, Welch began showing early signs of pregnancy, which workers at Auburn assumed was most likely due to sexual relations with a male guard. Rather than being released and given the medical attention she needed to carry out her pregnancy easily, she woke up every day, confined, with her maternal health unaddressed. But being deprived of proper accommodation was not her only problem. By the beginning of her third trimester, a guard, Ebenezer B. Cobb, flogged Welch. The visible and sustained marks on her body, as well as the mental weight of being incarcerated and pregnant, led to a decline in her health. In December 1825, Welch eventually gave birth to a child despite her confinement and physical abuse. Postpartum life can be rapacious to the body, no matter the environment, though that challenge was amplified during Welch's continued confinement at Auburn. Unsanitary conditions, poor sleep, and chilly weather terrorized the young woman during and after her pregnancy. However, one month later, on January 9, 1826, slightly over a year after her arrival, she was dead. Her death did not go unnoticed and prison authorities placed blame on the guard who lashed her. Eventually, Cobb faced a grand jury in connection with the whipping of Welch and its contribution to her ill health and death.

The jury acknowledged the harm he caused Welch, and Cobb was convicted of assault and fined $25. Despite the ruling, Cobb retained his job. When one thinks about the brutal structure of a nineteenth-century women's prison, there is enough evidence to see that the conditions were abject and inhumane. And even the abusers only faced a small punishment. Welch's case is one of many haunting episodes of acute isolation, sexual assault, and grisly death. In contrast to the lack of action at the abuse of female prisoners in

New York State during this time period, the controversy surrounding Welch's death led to mild prison reform.

After her death, New York State officials deliberated on how women should be incarcerated, marking a new epoch in state prisons. Rather than challenge the jury's decision for Cobb to retain his job, governor DeWitt Clinton objected to the condition of the prison itself, insisting that the predicament was "the lamentable condition of female convicts, and the manifest and gross impropriety of their ever being confined in the same prison with male convicts." The dilemma, for Clinton, was having male staff on the grounds, not the prison itself. This meant that the Auburn system, which had male and female prisoners at the same site, albeit segregated, was suddenly abrogated, and Clinton pushed the New York Legislature to authorize the building of Mount Pleasant Female Prison at Sing Sing grounds on the Hudson River. When it opened in 1839, Mount Pleasant was located in a two-story neo-classical house lined with Doric portico. Rather than confine the women in attics, as they were previously, inmates were placed in twenty-four cells. As the incarceration of women grew in New York State, the accommodations remained the same, to the point that Mount Pleasant faced overcrowding in the 1840s.

Unlike Auburn, Mount Pleasant abandoned the silent rule, and they restricted extreme physical punishment. Instead, they opted to provide literacy courses to inmates. Prison staff established a nursery to house the newborns and toddlers of inmates. While the conditions at Mount Pleasant were significantly better than those at Auburn State Prison, the building was overcrowded, women were malnourished, and custodial officers placed some inmates in straitjackets for extended periods. In 1877, the New York Board of Inspectors recommended shutting down Mount Pleasant. They sent some of the incarcerated women to Bellevue Penitentiary in New

York City, but eventually, state officials wanted a larger institution to detain women throughout the state. Rather than question whether a prison could rehabilitate women or be humane, they decided to create a new one. As a consequence, in 1901, they built Bedford Hills State Reformatory, which they believed would be an alternative to the harsh conditions at Mount Pleasant.

By the end of the twentieth century, women's prisons in New York State varied in their architecture and system. The traditional methods of punishment—silence, gagging, or straitjackets—were not officially permitted, but more than that, the incarcerated actively created something new.

When the ACE Program was established at Bedford Hills in 1988, there were many people who were incarcerated precisely because they did not conform to society. Many were queer, many were leftists, most were born in impoverished families. A segment of the inmates was incendiary by virtue of being incarcerated. What made this period distinct from the early nineteenth century was the degree that these women were collectively addressing the concerns of inmates who were also subjected to the new plague.

In 1988, forty women gathered to make a quilt. At first, they met in a mess hall in upstate New York, using everything they had—cloth, pencils, notes. Their labor was an act of care: to make a work of art to honor their deceased friends—fifteen women who died of HIV/AIDS at Bedford Hills. Most of the departed were African American and Latina women, mainly from the working class. Together, the surviving inmates hand-stitched the quilt, ultimately providing a different square for each deceased person. For the participants, the aesthetic exercise also freed them from their social isolation. Eventually, they put it on display in prison, and the names of the individuals honored were sent to the NAMES Project.

For people seeking community and closure, AIDS memorialization prompted a catharsis. The AIDS quilt, a fifty-four-ton panel tapestry, is one of the most well-known and revered community mementos of the twentieth century. By giving a name to the dead, with a visually colorful display that honors each perished AIDS victim with a unique square, the quilt is personalized and in effect it destigmatized the disease. In Bedford Hills, the prisoners' contributions to the quilt had profound influence for inmates who wanted a public way to grieve the loss of their friends who died of HIV/AIDS. "The quilts were laid out, fifteen of them, on the floor of the school building corridor. They stayed there for a week. Day after day, women passed through looking at the names, unable to turn away from the reality." Behind these prison walls, they were denied traditional funeral rites, but making the textile was a collective form of healing.

The late intellectual Susan Sontag once commented: "The sexual transmission of this illness, considered by most people as a calamity one brings on oneself, is judged more harshly than other means—especially since AIDS is understood as a disease not only of sexual excess but of perversity." Pathogens, Sontag argued, are biological but how we cope with them is political. She pointed out that our feelings about a disease might carry as much weight as a cold, cough, or headache. When the AIDS pandemic struck New York City, Sontag described how the epidemic engulfed her community and loved ones, the panoply of death also causing a depletion in queer talents and dreams. Like writing, the quilt was a way to honor the dead.

Prison Medicine Today

Prison continues to be a major incubator of disease. Even as health-care services expanded over time—as in New York State—the institutions lack the expertise and staffing to treat the broad sweep of medical needs presented by prisoner populations that can grow to a thousand or more at a single site. Dr. Rachael Bedard, a former geriatrics physician at Rikers Island Correctional Facility in New York City, found this out for herself in 2016. When she began working at Rikers, there was momentum to provide a fairer, safer experience to the prison's six thousand inmates. Bedard felt optimistic that she, in caring for some of the most vulnerable inmates in Rikers—older adults—could make a significant impact in her new role as physician and healthcare consultant to the facility.

When she joined the system, Bedard was encouraged by anti-prison social movements to reduce the number of incarcerated people in New York City. As a physician working within the prison system, she hoped to improve the health of inmates. For progressive healthcare workers, a consensus had seemed to seemed to have around the inherent risks of incarceration itself, as documented in Homer Venters's *Life and Death in Rikers Island,* where he stated, "Generally, there is a quick response to condemn an individual nurse or doctor for failing to do their job, but with no discussion of the pressures that led them and thousands of others to stray from the path of patient care." Health workers are drained from the incessant demands of treating confined patients.

Prisons, the data shows, are a place where death is accelerated. When I asked Bedard about her experiences at Rikers, she said it was common to see younger inmates—far younger than those aged fifty-five and older as her specialty prescribes—exhibiting geriatric syndromes. Their bodies had aged. What she noticed was that they had chronic illnesses such as hypertension, diabetes, and heart failure run amok, often outwardly concealed by the inactivity and seclusion of the prison setting. Limited as she was, Dr. Bedard and her geriatrics team could at least intervene by advocating for compassionate release orders for pretrial prisoners who faced debilitating symptoms.

Inmates do not only have to worry about premature aging—sexual health was also a lingering problem. When I spoke with Dr. Jaimie Meyer, an infectious disease physician, about what it is like to work with HIV-positive patients entering the penal system, she explained some of the barriers that incarcerated women and non-binary people experience are particularly harrowing especially given the particularities of reproductive and gender-affirming health. Noting, "providing screenings for cervical and breast cancer through pap tests and mammograms" allows the doctors to detect chronic diseases before they materialize. A challenge, she found, is that many of the prisoners have experienced trauma, and they don't want to go to a healthcare provider and disclose their whole life history, given that the doctor is still associated with the prison system.

The restraints are inescapable. In 2021, when Dr. Rachael Bedard left her job at Rikers, she did so out of frustration with the inertia she'd run up against. "The system feels sort of patently unsafe for my patients," Dr. Bedard told me. "I was no longer working in a context where I was making the difference and I had to wonder what I was doing there." She realized that no matter what she did as a healthcare provider—the prisoners would become debilitated.

Dr. Bedard is not alone. A similar exodus happened with another New York City–based doctor. Dr. Ross MacDonald, the former chief medical officer of Correctional Health Services at Rikers Correctional Facility, echoed Dr. Bedard's critique. To his mind, the structure of prison as it exists in the United States is incompatible with healing. Over the course of his decade-long tenure inside the correctional system, he told me he had seen the private healthcare system employed inside prisons swell into a profit-making enterprise above all else. This dovetailed with an expansion of the incarcerated and an incentivized lax treatment. "The public hospital system," he disclosed, "allowed us to have a group of clinicians who desired to see incarceration happen."

In 2021, the New York State Legislature finally passed a law that pushed private care companies out of the state's prisons and replaced them with an independent health authority, separate from the Department of Corrections and equipped with its own budget and organizational structure; clear progress that unfortunately remains far from the norm across the United States. Dr. MacDonald has seen new opportunities to recruit and employ clinicians aligned with a more humane vision of incarcerated healthcare, but the system is still burdened with pathos.

Physicians contracted to work in New York State and City prisons are constantly confronted with a penal system that restricts when and how their patients get the care and how they move. Dr. Bedard acknowledged that "people do not have the freedom or the latitude to be able to do anything for themselves." Whether they are heading to the cafeteria or visiting a family member, every time a prisoner moves, someone has to move them from one space to another.

Like the prisoners at Bedford Hills, these doctors questioned the prison as an institution. Dr. Bedard noted, "The jail setting is incompatible with providing community standard of care to people who are seriously ill. I think anybody who spends significant time in jail or prison knows that the environment is not rehabilitative, rather, it is so fundamentally traumatic that I came out." She continued, "I went into the work because I like it from a human rights–driven perspective. But I came out of the work like a much stronger abolitionist than I was when I came in." This abolitionist tone has spread to other healthcare workers in prisons. As doctors and researchers who worked with the incarcerated in the northeastern part of the United States, they shifted degrees of power primarily dictated by the penal legislation, the number of the detained, and the resources available to them.

Rather than see medicine as individually based, doctors such as Bedard, MacDonald, and Meyer believe that collective and ethically grounded approaches to treatment can challenge health inequities. Their message echoes among a new generation of physicians in America. As Dr. Eric Reinhart wrote, physicians who work in the United States need to develop a more profound vision of care: "the abolition of inherited structures of oppression that often subtly reproduce themselves in our very efforts to undo them." It is not enough to provide primary medical care—as *Estelle v. Gamble* guarantees—but, as cardiologist Dr. Eric Topol explains, "it's possible to imagine a new organization of doctors that has nothing to do with the business of medicine and everything to do with promoting the health of patients and adroitly confronting the transformational challenges that lie ahead for the medical profession." The problem then becomes, what happens to the incarcerated when there is an outbreak: Will they always have to fend for themselves?

Even the most well-meaning doctors have seen the limitations of their practice in prison, and because of their social status and profession, for the most part, they do not know what it is like to be both sick and incarcerated. Their compassion and testimonies, while illuminating, cannot take away from the history and continued set of literature produced by incarcerated people and the stakes that are taken to write about one's health from within the prison. Nevertheless, these healthcare workers are wrestling with the moral limits of working in prison, aware that being tethered to the prison means that no matter how well-meaning a physician is, they are implicated in a system predicated on expanding the carceral state.

While most conservatives and some liberals in the United States embrace the need for prisons, with some arguing that this is the only way that some of the inmates receive healthcare, the healers who work in these settings find that the physical and mental well-being of the incarcerated is compromised, especially during a pandemic. If we care for other humans—even those who might have caused harm—there is a compelling case that forced detention, especially given how it makes people susceptible to malady, might not be the solution.

In Their Own Words

I first encountered Dr. Angela Davis's work in a New York City reading group. With eyes opened wide by her *If They Come in the Morning,* a treatise on incarceration, I was blown away by a revolutionary feminist who could speak personally about the harms of prison. "By almost any standard the American prison betrays itself," she wrote, "as a system striving toward unmitigated totalitarianism." Davis recognized, like many radical philosophers before her, that the prison—in its role to make humans captive—could do this by imposing despotic rules on a confined population. Her text blended the personal with the political, in such a way that she heralded sympathy.

Davis's detention, a highly politicized and internationally noteworthy event, inspired letters from Berlin to New York. In "An Open

Letter to My Sister, Miss Angela Davis," written at the time of her confinement, the esteemed James Baldwin opined: "What has happened, it seems, and to put it far too simply, is that a whole new generation of people have assessed and absorbed their history, and, in that tremendous action, have freed themselves of it and will never be victims again." He continued, affirming that Black Americans had been taught to hate themselves, an outgrowth of the free-form subjugation and captivity of the country's Jim Crow era. But given the Civil Rights movement's initiative to overcome social adversity through direct action and education, Baldwin and other Black philosophers surmised that even in captivity, one can articulate one's personhood through writing. Prisoners who were physically detached from their families attested to the ways that confinement tightened their grip on the world, and with each other.

For the incarcerated people at Bedford Hills, who knew that freedom was out of reach, composing slashed through the confines of their forlorn state. More specifically with members of the ACE Program—writing was braided into their lives. They wrote illness narratives documenting their internalized stigma about HIV and their position on providing care with brutal honesty and clarity. *Breaking the Walls of Silence* is a sociological manifesto boldly proclaiming who they are—a group of people who strove to chronicle how they endured a system that was meant for them to rectify themselves. They were an enclave of women lodged in the affluent suburbs of Westchester County. But they were also mothers, partners, girlfriends, wives, and siblings who created prison medical literature centered on women. Still, it was not just a declaration about the HIV/AIDS epidemics but an archive for the incarcerated, deeply tied to radical roots. They sketched—from the corridors of prison—how they cared for people living with HIV, but how they also educated each other about what the disease meant. Through testimonials and poetry, they illustrated how they carried out their peer education program and how they used writing to muster the community within the prison walls.

People find community in organizing. In his book *The Dawn of Everything*, the late David Graeber contended, "When sovereignty first expands to become the general organizing principle of a society, it is by turning violence into kinship." This was evident in how ACE members imparted peer support, a quilt project, and a writing program. For the most part, ruthless systems such as a prison can and do degrade and deform people—yet collective acts that generate agency can provide sustained solidarity among prisoners. At Bedford Hills, the burden of incarceration and bodily disintegration meant that prisoners had to find care in a system that provided very little. Inmates abetted late-stage AIDS patients by bathing, feeding, and reading to them. In effect, they became surrogate families, in the high coolness of the seasons. Writing at the margins, during the beginning of the AIDS epidemic, was a sovereign push against the carceral state's tainted institution.

The power to tell one's story is seismic, and that is why the inmates at Bedford Correctional Facility formed the ACE Program. Although oppression creates and sustains pandemics, storytelling can be a way to bare one's soul and document how agony can be transcended. Unable to move freely beyond the walls, their words and their book could travel through the impeccably vertiginous bureaucracy of the prison state. Some people think prisoners should be made to wither and die alone in their cells. ACE showed that even when captive, when one is sick under the most brutal circumstances, one discover something beautiful within oneself, the power to provide care, even in a carceral state.

Abolition as a Form of Care

I was eight years old the first time I visited a prison. This trip—which would be the first of many—occurred after my aunt N was rescued from the Atlantic Ocean with nearly one hundred people. Her holding place was classified then and now as the Krome North Service Processing Center (or Krome). Twenty miles south of downtown Miami, it shared all the traits and emotions of a prison—a place for migrant confinement. US authorities interned undocumented migrants, restricting where they could go and whom they could see. The prisoners were uniformly dressed in orange jumpsuits, a jarring contrast to the visitors who entered this restricted space. Undeterred, my parents, aunts, cousins, and uncles visited Aunt N every weekend we could, bearing smiles and candy. I remember playing at the outdoor visitor's area at Krome with the other children whose relatives were detained. Although there were no toys for us—restless, we ran in the courtyard.

When a country, like the United States, commits itself to building more prisons, this is an indication of the value society places on particular groups. The people we visited were not guilty of causing any physical or psychological harm to others. No one at Krome had seized or stolen a piece of property of any citizen of the country that now held them captive. These were Haitian migrants seeking asylum, and Dominicans seeking economic lift in the place so often marketed as the "land of opportunity." Rather than immediately processing and releasing Aunt N—or anyone else who was detained—the state believed it best to deny her freedom for eight months. This was the first time that I became aware of what prisons meant. They were not a place to maintain a social order, as I had been taught in elementary school, but rather a tool to actively separate families.

Aunt N, like other relatives who migrated from Haiti, became temporary fixtures of my home. She became the sixth person to move into our two-bedroom pink duplex in Little Haiti. Our living quarters grew into a cramped space, where each person dreamed of privacy. Although Aunt N was free, the US government did very little to help her transition. Instead, my parents, some other elders, and the community writ large scooped through the few resources they had. Eventually, she found a job and started learning English at an adult after-school program. For me, Aunt N was a success story because she had community. But immigrating to the United States since the mid-1990s has significantly changed. Later, I learned that women of all backgrounds—migrant, poor, and Black—would find themselves behind bars. "The convergence of local politics, state laws, institutional policies, and law enforcement practices criminalizes unauthorized immigrants," argues scholar Amada Armenta, "and deposits them into an expanding federal deportation system." That program, according to Armenta, is not race neutral nor is it tranquil; rather, the process brushes against an intensely visible carceral state.

Beyond the emotional toll of witnessing a detained family member, there is a cost when a child visits an interned relative, being unable to hug them or hear them sing. Aunt N was eventually manumitted; her time at the Krome North Service Processing Center robbed her of her dignity and time.

Some theorists provoke us to examine the structures that cultural norms train us to believe are crucial to the maintenance of social order; others make us reflect on what it would mean to envisage a world that provides some repair for the incarcerated. Others go further and raise the question of what we truly believe will result from the incarceration of a fellow human.

One of the most profound sociologists on prisons is Ruth Wilson Gilmore. Her work germinated from unearthing a problem: "the

BREAKING THE WALLS OF SILENCE

phenomenal growth of California's state prison system since 1982." Published in 2007, *Golden Gulag* traced the intricate and wanton patterns in policing and prisons. What Wilson Gilmore found was that working-class people, especially ethnic minorities, were more likely to be seized and incarcerated. And while many of them lived in communities that needed more school funding and clinics, the prison budget in the state of California kept increasing. So, when she carefully read and collected the data, and visited detention centers throughout her state, she reached one conclusion: that prisons must be abolished.

A younger generation of scholars have augmented Wilson Gilmore's claim, also showing how mass incarceration in California was not an anomaly. In her book *Carceral Capitalism*, the scholar and prison abolitionist Jackie Wang demonstrates how Florida—her state of birth and mine—had policies similar to those of California. They started divesting from social programs in the mid-1990s while simultaneously expanding the prison system. Wang asked her readers to question not only the legitimacy of prisons, but the concept of innocence and criminality in the first place. Exercising doubt might cut into the fabric of American society, but given the major gulf between the intention of prison (rehabilitation) and its reality (maltreatment), it is difficult to make a moral case for prisons.

From the time Rachel Welch was detained at Auburn in 1825 until Kathy Boudin was incarcerated at Bedford Hills in 1984, prisons in New York have gone through drastic change. For one, the number of incarcerated women has augmented. Between 1980 and 2020, the number of incarcerated women increased by nearly 500 percent, rising from 26,326 in 1980 to 152,854 in 2020. At times, prisons have been a site to place our wayward kin and at others to constrain political radicals. But what has been especially grave is that many of the women who are incarcerated need psychological

135

care and safety. A 1999 study at Bedford Hills found that 94 percent of the women interviewed had experienced physical or sexual violence. Over half of the women in New York prisons (52 percent) have been diagnosed with mental health conditions by the Office of Mental Health. Every seven days, an incarcerated woman self-harms or commits suicide in a New York State prison. While many correctly cite mental illness as the cause of these gruesome figures, the extreme manifestations of mental health conditions are at least partly due to the ills of incarceration. Liberals, who are sympathetic to these figures, advocate for small prison reforms. But anyone who has had direct contact with an incarcerated loved one knows that more needs to be done.

When "safety" is the only goal, we get unmoored solutions that are perhaps safer, or perhaps look like reform, but they fail to acknowledge how the prison system is invested in pervasive purgatory. As Maya Schenwar and Victoria Law remarked in *Prison by Any Other Name*, "reform is not the building of something new. It is re-forming the system in its image, using the same raw materials: white supremacy, a history of oppression, and a tool kit whose main contents are confinement, isolation, surveillance, and punishment." Even when people have the best of intentions, prisons cannot ever be safe; rather, they kindle illness.

Debates on how, when, and for how long we should imprison our fellow man for antisocial behavior will rage perhaps forever—but what cannot be up for debate is the need for prisons to be free from torture, equipped to supply the basic needs of healthful living, and in the barest sense, outfitted with the staff and equipment to prevent them from becoming vectors of internal and community spread of vicious diseases, as they once were when the ACE Program was established. Regardless of what we think of Kathy Boudin, Katrina Haslip, and Judith Clark, they established a framework to care for some of the most marginalized women in society—HIV-positive incarcerated women. Their efforts, on the surface, might be seen as sheer prison reform. But what it shows is that if these prisoners were free, they would have been healthier.

In the summer of 2020, shortly after George Floyd was killed by Minneapolis law enforcement, many people questioned the root cause of his murder—America's police state. First, people occupied their neighborhoods, choked with rage and grief, but also an insistence that something needed to be done. Others found themselves turning their attention to the overwhelming 1.5 million people who were incarcerated. Abolition hurled through their mouths, unleashing what people thought would be a new era. But over time, some of us were disillusioned and referred to this moment as the racial reckoning that wasn't. A world without prisons has become a far-off possibility in today's political environment, but the millions of people residing in prisons know that the only possibility for their flesh and spirit to be robust is for them to be free.

Chapter 5

EBOLA TOWN

We can only understand the present by continually recurring to and studying that past; when any one of the intricate phenomena of our daily life puzzles us; when there arise religious problems, political problems, race problems, we must always remember that while their solution lies in the present, their cause and their explanation lie in the past.
—W. E. B. Du Bois, "The Beginning of Slavery,"
The Voice of the Negro, 1905

With medicine we come to one of the most tragic features of the colonial situation.
—Frantz Fanon, *A Dying Colonialism*, 1959

I N 2014, DR. PHILIP IRELAND saw flashes of lightning before his eyes. What began as a mild headache turned into a thunderous roar in his skull. No matter his doubt, he knew his affliction wasn't typical, and that something far graver was wrong. Like any doctor worth his salt, he was fastidious about his diagnosis, so he took his temperature, in search of more clarity about his ailment. With a temperature of 101 degrees Fahrenheit, Dr. Ireland initially thought he had malaria, but given that he had had the malady countless times before, these symptoms felt different. His body was sluggish and suffered from a peculiar sensation in his right hand. As the day progressed, his worst fears about the pathogen attacking his body moved closer and closer to reality and he believed that his symptoms "sounds like, smells like Ebola."

That summer, Dr. Ireland was among the many patients who felt Ebola's wrath. The viral hemorrhagic fever was spreading, far beyond the African continent and transmitted via boats and airplanes by people like Dr. Ireland, who had come into close contact with the bodily fluids of Ebola patients. Untreated, the disease has a 60 percent fatality rate, but before death it exhausts the body, shutting down the normal functions of the liver, kidneys, and brain. The "flash of lightning" Dr. Ireland described paints a picture of how the virus pushed his body to its ends as it launched an outright war.

During the height of the 2014 outbreak, the Zaire ebolavirus was rampant and lethal. In some patients, Ebola leaves the skin withered, the body trembling and listless. Whether someone is experiencing a fever or nausea or a purulent discharge from the anus, the ailment unsteadies the body and mind. Often people infected with Ebola report a lack of appetite, difficulty breathing, sore throat, and weakness. Initially, the symptoms appear flu-like, but internal bleeding can contribute to organ failure. The human cost to pathogens is far more critical than the symptoms themselves.

One of the most significant factors affecting survival is what they lose. That is to say, dehydration, often caused by the loss of urine, blood, and sweat, can leave a person needing fluids replaced. Other symptoms, such as joint pain, reddening of the eyes, and mucosal bleeding, can occur for as long as twenty-one days. Like cholera patients who expel a large amount of fluid, providing Ebola patients with replacement fluids can be essential to saving a life.

Survival is at the heart of the significance of Ebola; how people live or die unearths the core of what it means to overcome an illness.

For the first several months of the Ebola outbreak in Liberia,

Dr. Ireland had no choice but to diagnose, isolate, and treat Ebola patients who came to his hospital. As an emergency medicine physician at the John F. Kennedy Memorial Medical Center in Monrovia—Liberia's largest hospital—he was expected to uphold his medical duties to treat the urban population of this capital city even if the disease was difficult to manage and contain. Born, raised, and educated in Liberia, Dr. Ireland had ample experience with infectious diseases, and knew that early diagnosis and medication could stave off most illnesses. He, like many other doctors in Liberia during the early days of the 2014 epidemic, worked alongside the Ministry of Health and the World Health Organization to assist with Ebola efforts. Yet, despite ample experience dodging infection, he, too, became a victim of the virus.

Trying not to overindulge his fear, Dr. Ireland initially contacted the Chief Medical Officer of his hospital, informed him about his potential exposure and growing symptoms, and asked to be tested. Shortly after, he was isolated for twenty-four hours as he awaited his results. Knowing about the biology of transmission, Dr. Ireland secluded himself and instructed his mother on how to provide him socially distant care. After his recovery, he wrote in an editorial, "My mother, like thousands of other Liberians, was on her own when she saved my life because she had no other choice." Like many societies, Liberia relies on the unpaid but essential work of women's care. Liberia benefited from this domestic work, yet given the gravity of the disease, the entire medical system and its staff were vulnerable during the early stages of the epidemic.

Healthcare workers of all stripes contracted Ebola that summer, and some were not as lucky as Dr. Ireland. Samuel Brisbane, the emergency department director of the John F. Kennedy Memorial Medical Center, died in July 2014. Grieving the loss of their friend and colleague, physicians Josh Mugele and Chad Priest sought to treat Dr. Brisbane's death with the honor it deserved, a reverence for healthcare work rarely seen until the nightly remembrances that marked the first weeks of Covid-19 in 2020: "For emergency-medicine clinicians like us, the concept of a good death can seem too abstract, intangible," they wrote in the *New England Journal of Medicine*, "rarely are the deaths we see good or beneficial."

Given that a significant number of healthcare providers passed away during this critical period, the researchers surmised that their passing echoed the ancient conception of dying for one's homeland. Their account braids together the nobility of healthcare workers and the inevitability of death—which were especially apparent during the first several months of the outbreak. What was striking about what happened in Liberia, however, is just how limited frontline medical workers were in their ability to deliver optimal care without the benefit of the essential medical supplies required to stanch Ebola's transmission. While his colleagues' memorialization of Dr. Brisbane underscored the ugly and complex realities of this deadly virus, measures existed that could have saved his life and others'.

Healthcare workers often have to step outside a scenario— thinking through the lesions that bleed their society. The lack of funds, the dearth of staff, a shortage of materials and equipment. "The sick continue to be turned away, only to return home and spread the virus among loved ones and neighbors," expressed Dr. Joanne Liu, then-president of Doctors Without Borders. She later made a case for a prurient public health plan that would include isolation centers for people who chose to receive remedy. But bestowing medical care to Ebola patients was not the only problem; other health issues such as cancer and tuberculosis were neglected, leaving already vulnerable groups in an even more dire situation. When Dr. Liu was in Liberia, many healthcare facilities were closed by mid-August 2014.

The Liberian medical system, already stretched thin by an unfathomable plague, was dealing with a crisis of resources and services. This also left pregnant people and malaria patients unable to get necessary provisions.

Very few diseases create such supine patients as the Ebola virus. Similar to Dr. Ireland, other people had excruciating headaches and erratic visions. For the first several months of the outbreak, established hospitals such as the JFK Memorial Center and makeshift hospitals supplied as much staff as possible. Every biohazard suit, mask, and public health worker was used to contain the contagion, but they could not cast off the entire epidemic.

While many well-intended physicians and caregivers worked on relief efforts, their campaigns sometimes lacked cohesion. That year, the World Health Organization noted, was "the first time the disease has been detected in West Africa and the outbreak has now spread to the American and European continents. More than 22 million people are living in areas where active Ebola transmission has been reported." A conglomerate of public health entities funded efforts—the Liberian government, the World Health Organization (WHO), and a patchwork of non-governmental organizations—initially fumbled through the novelty of the situation when they rolled out their resources. The World Health Organization, then responding to other concurrent outbreaks—MERS in the Middle East and Avian flu in China—was overwhelmed by the litany of outbreaks in 2014. Though these other pathogens fractured the political response, the more significant issue was financial. Moreover, the WHO African Regional Division budget had been halved from $26 million in 2010 to $11 million in 2014. Having eliminated staff and resources, then-director Dr. Luis G. Sambo cast doubt on the division. One temporary salvation was Ebola's interlude.

After a lull in Ebola cases in West Africa in the early summer of

2014, the virus began to spread in the Liberian capital of Monrovia in July, which meant that public health measures came to a standstill. But that did not come with political stability. On August 7, Ellen Johnson Sirleaf, then-president of Liberia, declared a state of emergency, proclaiming "extraordinary measures for the very survival of our state and the protection of the lives of our people." In practice, this meant that she rescinded civil liberties, saying that "ignorance and poverty, as well as entrenched religious and cultural practices, continue to exacerbate the spread of the disease." She closed the borders and asked the Liberian army to monitor and occupy parts of the country. President Sirleaf surmised, at the time, that the health of the nation would come at a cost. Instead of increasing the number of doctors, which she believed was impossible, or giving out more protective medical gear, which the Ministry of Health did not have, she believed in circumscribing a subclass of people in the capital city.

The way a leader responds to a health crisis has a lot to do with how much that person trusts their population. There is something profound about how a ruler responds, as it holds a mirror up to society. "Capital is reckless of the health or length of life of the laborer," contended Karl Marx, "unless under the compulsion of society." When he wrote this, Marx reflected on the state of industrial workers who churned through ten-hour workdays, occasionally slipping into states of bodily fatigue and occupational injury. Yet his introspection about the fragile condition of health under capitalism emphasizes something relevant about people, whether in Liberia or elsewhere, if society could be gentler to the worker—and the poor. There are things that public health measures afford, the ability to address mass infection and death through a simple fix, given that the field is constantly changing. In the face of technological and medical innovation, and social pressures, politicians have to work through ways to heal the nation from the microbes that invade a population. Nevertheless, the delicate balance between health and civil liberties shows how intractable health problems can be. President Sirleaf's plans to lock down a country rested on a belief that Liberians had to be contained, given the material failures of the state. The contagion cut through the fault lines of Liberian society.

Brimming with an uptick of cases, the Ministry of Health built more clinics during the middle of the outbreak, which to their surprise were challenging to sustain, given the death toll of health workers. Medical centers were hastily constructed to compensate for the lack of space available at the three leading hospitals in Monrovia. Nevertheless, there were fearless healthcare workers within the health system, notably Dr. Mosoka Fallah, a Liberian infectious disease researcher, who worked to mobilize contact tracers in Monrovia. As a public health expert, he took a holistic approach to work with low-income communities, not only attending to the sick—without infecting the healthy—but also finding ways to bury the dead with less risk. But this work took a toll on Dr. Fallah and others. Slowly, they built Ebola treatment centers that became an anchor for isolation, yet scarcity, he told *TIME* magazine in 2014, caused a real strain:

> August and July were quite tough for us. People would die, and we would be helpless. We just couldn't do anything. I would have a contact tracer follow a family. The mother died. The sister died. The maid died. The wife died. The father died. And she [the contact tracer] would go there every day to do contact tracing on the symptomatic people.

His account brings the horror close to home that, even with intervention, people could lose everything.

Throughout West Africa, Ebola isolation centers were established early during the outbreak in 2014. Doctors Without Borders—which had long-standing ties to the communities in the region—used their connections to establish Ebola isolation centers near Monrovia. As public health officials tried to close the gap between resources and facilities in the ensuing months, they drew plans to construct more ad hoc centers. Without informing the residents, the Liberian Ministry of Health converted a primary school into a holding center for Ebola patients in West Point, a working poor community in Monrovia. Initially, this was meant to supplement the well-worn establishments that could not isolate every Ebola patient. Still, the center's presence led to a deep divide between health authorities and the community. Bewildered, the residents were distraught that they had little input into these decisions or knowledge about who was being put into the facility and were upset that the government was importing Ebola into their community. One resident, Christiana Williams, then fifty-two, recalled hearing cries from Ebola patients at the makeshift clinic. Other residents were frustrated about this novel disease—because they did not understand the potential threat. Instead of harboring trust, the policy cradled skepticism in West Point.

When I spoke to Tolbert Nyenswah, who was then the Assistant Minister of Health in Liberia, he reflected on the positive aspects of their work.

We had a unit for psychosocial support, media engagement, and laboratory work. Our somewhat decentralized to Ebola response was initially a top-down approach. This meant we had a team that was working with community-based organizations and civil society organizations before the international support started coming to Liberia.

Nevertheless, the quarantine of West Point did not bode well. The community was agitated by the Ebola treatment center being placed near their homes. Some protesters targeted the Ebola isolation station and others believed the Ebola virus was a fabrication.

The word "lockdown," in all of its variations, has been laden with heavy meaning since 2020—still, the essence of the term is to be bludgeoned with captivity. During the 2014 Ebola outbreak, containment became a central theme of how Liberians referenced Ebola abroad. Successful public health measures rely on cooperation. Though for those who can move, remaining still can be unbearable. When one is forced to quarantine, one has to believe that confinement will aid them and that they can survive persistent immobility, especially when they haven't been infected with a microbe. When Ebola became an international concern, President Barack Obama insisted that the illness "can be controlled and contained very effectively if we use the right protocols." With over six hundred confirmed Ebola-related deaths in Liberia by August 2014, politicians on both sides of the Atlantic Ocean were trying to work through how to prevent a surge in cases and mortality.

On August 20, 2014, President Sirleaf imposed a twenty-one-day quarantine on West Point, which left many people unable to work or leave the community. Residents, objecting to the president's measures, believed they were in an open-air prison. This decision to quarantine a neighborhood and create a state of emergency was particularly unsettling, leaving some humanitarian workers to refer to the *cordon sanitaire* as a "plague village." While the politics of health and care do not always map onto war and its metaphors, public health is messy when new and virulent diseases take over a country in bereavement.

Lockdown City

At a cursory look, videos of West Point during the lockdown were damning. In one clip, my eyes fixated on the weather: the gray sky and heavy rain. Except for armed police, the streets were mostly empty. Some fruit stands were overturned. In another film, a young man stood behind green bars with several people near him. Solemn, he stared straight at the camera remarking: "We expect the government to come out with awareness; that's what we expected. But at four o'clock in the morning, they deployed police, army, and immigration. They beat people, and that's not right." Disillusionment arises when the state turns against its citizens.

As West Point residents were pottering about their errands, their lives were halted by a public health system that wanted to contain the disease. For some, the lockdown was sprinkled with physical terror as questions arose about where to place Ebola victims. The Liberian government's blockade began when the military prevented people from the western regions of Grand Cape Mount from entering the capital city. For many, the military siege meant street vendors could not sell their produce at the markets or provide people with staple goods. An alternative to this approach, one that would have greatly benefited this community, would have been a nuanced public health response that financially supported the people who were unable to work because of lockdown measures. Alternatively, providing free mental health services would have eliminated the isolation that some citizens felt.

Bordering the Atlantic Ocean, the West Point neighborhood is home to over seventy thousand people. During an epidemic that otherwise made life hectic and clamorous, the militarized languor of the residents was glaring precisely because Liberians who were already displaced from the recent Civil War felt abandoned by the government. They were reminded, once again, how little control they had.

The quarantine was not based on an honor system; instead, the Ebola Task Force was responsible for enforcing a quarantine as well as a curfew. The forced lockdown, for which the police constructed roadblocks and barricades, briefly tempered the population but as time went on they grew agitated. Soldiers and police officers in riot gear blocked the roads, and the waterfront was cordoned off, with the coast guard stopping residents from setting out in canoes.

For a fruit seller, the twenty-one-day incubation period would mean twenty-one days without wages and without government financial support. Loss of income poses a danger to a family's survival. By late 2014, Ebola had wreaked havoc on all sectors of society, according to the World Bank, leaving half of the Liberian workers without a job. The effect was ubiquitous—impacting food, dignity, and mobility.

Residents of West Point were shrouded with an active military occupation; they spoke of the ways they were made vulnerable, by being made subject to physical violence and food shortages. In the days that followed, some inhabitants were cut off from food and other essential subsistence. Davidette Wilson, a twenty-seven-year-old man, remarked, "There is nowhere to go for our daily bread." He pointed to his exasperation about the lack of sustenance. Bread, a basic necessity for life, was scant. Wilson's account showed the atrocious side of a lockdown, the close-ordered approach that left some residents in a situation where food was sacrificed for the possibility of spreading contagion. Beyond that, he also contended with the cruelty and inherent violence of being unable to subsist in a state of militarized quarantine.

Another chilling report revealed people's severed connections with their networks who lived outside of West Point. Patrick Wesseh, another West Point resident, asserted in a critique of the forced lockdown: "It is inhumane. They can't suddenly lock us up without any warning. How are our children going to eat?" For people, including Wilson and Wesseh, the quarantine was an unsettling experience that not only set up a physical barrier between the neighborhood and the rest of the city but outlined how they could easily be cut from essential services—the military could circumscribe the poor.

Among the people living in this neighborhood was David Anan. Then thirty-four, he asked gun-carrying Liberian soldiers, "You fight Ebola with arms?" For many Africans, living in a postcolonial context, militarization has been prioritized over care or harm reduction methods. Liberians, like Anan, were profoundly disturbed by the deluge of arms.

Knowing the physical and mental cost of confinement, the consensus was that the quarantine caused more harm than good. When the Liberian army met residents, the West Point community navigated in the best way they could—they revolted. Some people hurled rocks and stormed barbed-wire barricades, trying to break out, and others shouted as much as they could. For days, they filled the streets, which felt worthwhile given that very little autonomy or community discussion had been organized by the government. Unable or unwilling to address their concerns in a humane way, the government decided to take action.

On August 20, 2014, a barrage of military men, mainly from the Liberian army, entered West Point and shot live ammunition at a crowd protesting the lockdown. In the carnage, there were grim images, clearly indicating the explicit hostility of state repression. These snapshots confront the viewer with a fiery if not existential crisis. The armed men invaded the neighborhood, expecting coercion.

They intuited that punishment—through ordnance and terror—would calm the populace. What they witnessed was tumult.

After ten days of violence, a teenager was killed, leading community members to mourn. Grotesque as it was, the quarantine cast a shadow on how West Point residents were seen. Among the most powerless in the country, they were confined to this space, while white and non-African people who contracted Ebola were evacuated from Liberia. This was an indication that the public health assessment and the response were predicated not only on science but on what scholar Adia Benton has described as an "assessment about the value of life." That government's response is an indication of how little working poor Africans are valued.

Lockdowns, such as the one at West Point, are not evidence-based. They are an institutional means of controlling a group that has little dominion. When the quarantine ended, it was not merely about this neighborhood, or Liberia, or West Africa—but the fiction that poor African lives can be confined and disrupted without fair compensation. This wasn't unique to Liberia; the lockdown was applied to other nations in the region. Eventually, several Global North countries restricted travelers from Liberia, Guinea, and Sierra Leone, which led to political isolation.

The quarantine became the central point of dissent, not just because it restricted African travel but because there was a difference between how low-income West Africans were managed in the outbreak. Journalist Clair MacDougall contended, "The government exodus contributes to a sense among citizens that Liberia's wealthy and powerful have left the country's poor to fend for themselves. Many expatriates who work for non-governmental organizations and international companies have been evacuated, and their lavish apartments with 24-hour electricity and running water are now empty." MacDougall questioned the wayward politics and the double standard that exist in Liberia but on a global scale. Who is afforded quality healthcare and the ability to move and who is told to isolate, without funds, travel, or a livelihood?

The 2014 outbreak in Liberia showed that quarantine was not enough. Infectious disease specialist Amesh Adalja wrote, "While decisive action is needed to combat Ebola and other diseases, the quarantine of a geographic location applied to a people without evidence of infection functions not to control but to promote its further development." Besides lacking scientific credibility or moral standing, quarantining West Point assumed that all seventy thousand residents at West Point were incubating Ebola. To understand the outrage and the stakes of the matter, one has to consider how pre-existing social structures amplify inequities.

The lockdown was not just an inward event that trapped a community into a hostile environment; it was heavy, overwhelming, an example of what the late philosopher Lauren Berlant called "slow death." The lives of poor Liberians were temporarily suspended, bracketed by confinement, while those with far more resources could transcend it.

By August 2014, the WHO considered West Africa—not Ebola—a Public Health Emergency of International Concern. That is to say, the region, rather than the disease itself, was considered a public health risk, even though, by that time, there were cases in Italy, Spain, the UK, and the United States. The preliminary mismanagement of the Ebola virus outbreak and cordoning of West Point conflated militarization with health but did little to salve the cuts that wounded Liberian society. During the brief stretch of time that they were enacted, the policies had little to do with the life cycle of the microbe. If public health officials had looked at historical examples, they might have been able to provide more pointed and effective measures.

Ebola Uncovered

Before Dr. Peter Piot was stationed in Central Africa in the 1970s, he worked in an Antwerp laboratory that analyzed age-old infectious diseases. When he received a microbial sample on September 29, 1976, he and his colleagues were perplexed by "a nasty new contagious disease in some distant, benighted place." For Piot, the virus was a seismic shift. From what he was familiar with, this was a pathogenic malady more potent than yellow fever. At first, he wanted "new solutions that would save lives." So, he went to the virus's abode— Yambuku in the Democratic Republic of Congo.

Early accounts from Flemish missionaries and doctors in the Congo broke down what they saw: patients who were bleeding from their noses, exhibiting high fevers, and complaining of cataclysmic headaches. Some were comatose, barely hanging on to life. Piot wasn't trained in the field, but his desire to decipher and cure the novel virus drove his curiosity. In his recollection of the journey three decades later, Piot noted the self-imposed measures that various communities enacted to stave off the virus, noting: "About half of the villages had erected barriers to control people's movements in this time of quarantine." Protecting oneself from danger is a normal part of life, and when these people witnessed their neighbors drained of their vitality, they did whatever it took to survive. And while Piot's enthusiasm was notable, he was not the only researcher inquiring into the novel virus.

Congolese microbiologists such as Dr. Jean-Jacques Muyembe-

Tamfum were early witnesses to the 1976 Ebola outbreak, and were also perplexed by the microbe. When Dr. Muyembe-Tamfum went to Yambuku, he was tasked by the Minister of Health to "go there and assess the situation." At first, he thought the outbreak was typhoid fever. As villagers were dying en masse, he traveled to the village, equally interested in understanding the spike in deaths and treating patients with the few resources he had. When Dr. Muyembe-Tamfum arrived, he was astonished to find that a large portion of medical staff were absent from the Yambuku hospital. But from those who remained, he collected blood samples, noting: "I was immediately struck by the fact that when I removed the syringe from people's arms, the site of the puncture wounds bled profusely. My fingers and hands were soiled with blood." Dr. Muyembe-Tamfum eventually partnered with European doctors, who were stationed in the small town, to diagnose the outbreak. Although Dr. Jean-Jacques Muyembe-Tamfum was the first microbiologist to see Ebola patients and the virus's life progression firsthand in 1976, several months after he took the first blood samples, Dr. Piot and Dr. Guido van der Groen were credited with discovering the Ebola virus. Realizing that the fever needed to be understood beyond what the eye could see, Piot began harvesting the microbe to see how the disease propagated among survivors.

For scientists such as Dr. Piot, part of what made Ebola so compelling was its oddity, a feature that left researchers feeling exhilarated. When Ebola first emerged, few tools were available to analyze the physiological basis for this hemorrhagic fever: There was no genetic sequencing or effective treatment to combat the disease. As much as Dr. Piot tried to work directly with patients by observing their symptoms and providing direct care, he found that the high mortality rate made it difficult to document disease progression. The task, then as now, was to prevent disease transmission from the sick to the healthy. That meant providing people with a safe, comfortable space to rest as they replenished their bodies of depleted liquids and nutrients.

There are many origin stories for how an epidemic comes to fruition. In some cases, the virus mutates; in other cases, it jumps from an animal host to a human carrier. While pathogens with zoonotic origins are often misunderstood, they typically emerge simply because the animal in question, whether it is a pet cat or a small insect, live in proximity to humans. The scholar Donna Haraway describes this intimate relationship, noting, "Movements for animal rights are not irrational denials of human uniqueness; they are a clear-sighted recognition of connection across the discredited breach of nature and culture." Without sustained nuance, it's impossible to untangle the misconceptions about Ebola's origins, namely the idea that the disease has more to do with supposed African primitivity than with the universal reality that humans have close and reciprocal relationships with animals of all sorts.

People disclosed their Ebola accounts with vivid details like many novel infections. Initially, scientists were marooned with little knowledge of the illness. The perception of the disease to an outsider reads very differently than its effect on the patient. The threshold gap, albeit far-ranging, can be both oblique and harrowing. When researchers try to unpack a virus's physical structure, they hope its codes will give insight into its power. For molecular biologists, the dictum that genotype influences phenotype usually refers to fully formed beings, belying the idea that our genetic material and environmental factors shape our hair color, height, or eye shape. While far less complex than mammals, viruses also have observable characteristics whose genetic information can reveal the traits they express. The process of cracking Ebola's code has shaped research into understanding the evolution of organisms as we know them. By

2013, the genome of nearly thirty strains for ebolaviruses were sequenced, giving scientists further insight about their resilience. One of the fascinating biological aspects of this inquiry is that scientific research visualizes a tightly constructed picture of the virus in an effort to understand why and how the pathogen is destructive.

The Afterlives of Colonialization

History, even the national myth we construct, often shapes the power to respond and rebuild after devastating events. The 2014 Ebola outbreak in Liberia, Guinea, and Sierra Leone unlocked the vulnerability of postcolonial African societies. Epidemics do not exist solely as biological ailments but reflect ongoing social issues, and much of this crisis was deeply rooted in history that was forlorn. West African epidemics cannot be understood outside of the region's relationship to Europe and North America, and not only within the context of imperialism. The persistent, albeit tangled roots of economic extraction and political influence show that there are different shades of North American and European influence on West African shores. The pointed and critical moments of liberation did not remove the style of substance of colonization.

Liberia—whose history is intimately tied to the nineteenth-century coalition the American Colonization Society (ACS)—secured a novel "home" for the formerly enslaved people of African descent, who were forced to work and live in the United States and the Caribbean. The premise of ACS was that free Black people could "go back to Africa" and live in their "homeland." While it is not clear that these people came from present-day Liberia, African American and Afro-Caribbean people, who were sponsored to migrate to the territory, quickly found themselves in a tense and controversial society that neither invited them nor considered them kindred. The recently manumitted Black Americans were not part of the ethnic groups in the country, nor did they speak any of the indigenous languages. And because a few of them were given the resources by Western powers to colonize and rule over the indigenous Africans, their settlement led to a century of mutual resentment.

Joseph Jenkins Roberts, the inaugural president, was a wealthy free African American who settled in Liberia. During his political reign, some of the indigenous people were converted into Christianity and the intra-African slave trade continued to occur. This led to political tensions between Black Americans who established colonial settlements and indigenous Africans who were often displaced from their land. But alongside this deep restlessness, the power held by the former materialized into dominion over a small elite. In 1847, Liberia established its independence, with the country claiming itself as the oldest continuous African republic. Liberia's history is exceptional compared to its neighboring countries precisely because it was relatively immune from explicit European rule due to its status as a republic. When European countries met in the 1884 Berlin conference, popularly known as the "Scramble for Africa," officially, Liberia remained independent. Unofficially, the Black American elite did not operate alone. For most of its history, Liberia was influenced by a pastiche of external agents. While the United States has not been involved in every aspect of Liberia's internal disputes, there is plenty to show otherwise. They entailed in establishing critical institutions. Whether it was the Land Lease program, where Liberians aided the United States in providing arms to the UK, or the United States building two airports in Liberia, American foreign policy interests have shaped the country's infrastructure and military.

As Stephen Ellis commented in *The Mask of Anarchy,* internal strife in Liberia was not the fault of the United States alone. In some cases, families with roots tracing back to African American settlers established laws to force indigenous Liberians to work for multinational corporations. For example, the hut tax of 1925 extracted funds from the tribal population and forced many of them to work for companies such as Firestone Tire Company. Ellis notes that this meant that over the nearly thirty-year rule of William Tubman, from 1944 to 1971, his government enacted policies that utilized "income generated by foreign concessions to build Africa's first party-state at a period when other West African countries had not yet achieved independence from colonial rule." This open-door policy between the descendants of US-supported settlers and the indigenous continued to fuel deep divisions that had been festering for over a century.

The specter of conflict eventually came to the fore, leading to a military coup and two civil wars—one between 1989 and 1996 and the second between 1999 and 2003. During the Civil War, the healthcare system was ravaged, leaving most facilities unable to provide primary care for most of the population. Like any nation where working-class citizens are exhausted by the petrification of social hierarchies marked by origins rather than merit, people had to find calm in the afterlives of the military strife. Civil War in Liberia and the neighboring country of Sierra Leone meant that millions were forced into refugee camps, often living near the periphery of the cities.

When President Ellen Johnson Sirleaf won the election of Liberia in 2005, she eventually became the first woman president of an African nation after she was sworn in on January 16, 2006. However, that had little impact on the political factionalism or financial debt the country was moored in. Following a decade after the Civil War, internal political conflict and budget cuts and increased security ensued. The president was investing more in military and policing, and public health had policing incorporated into it.

At the helm of austerity and political distrust, the new government tried to hold the perpetrators of violence accountable, and there was an attempt to heal the nation. Coalitions became a way to assuage political trauma. In practice, they established a Truth and Reconciliation Commission in 2005 to diagnose and address the moral offenses that occurred during war. While this offered survivors of domestic war a platform to plead and respond to the past, the Liberians needed far more soothing than these political trials could provide.

After the Civil War, the pillars of health—and other social services—needed an antidote in Liberia. The onerous task of providing people with access to quality maternal care or cancer treatment was patchy primarily because the medical system was at its bare bones. In 2007, the Liberian Ministry of Health and Social Welfare spent $100.5 million on healthcare, which was 10.8 percent of its GDP, but this was not enough to build more hospitals and clinics in the cities and countryside. By 2010, several years before the Ebola outbreak, the country ranked 162 out of 169 on the Human Development Index, meaning that life expectancy and income were among the worst globally. Statistics can pass through our consciousness as markers of who lives and dies, but the burning concern is not how people fare during normalcy but how humans cope with the exceptional and the moments of mass contagion.

Tracking the particular experience of Ebola in Liberia is not a state of exception, nor is it an event that can be seen as mainly African; instead, it is an insight into how all people are invariably observers of epidemics. In his seminal book *The Plague,* Albert Camus wrote about the recurrence of epidemics:

Everybody knows that pestilences have a way of recurring in the world, yet somehow we find it hard to believe in ones that crash down on our heads from a blue sky. There have been as many plagues as wars in history, yet always plagues and wars take people equally by surprise.

When Camus wrote this, his chief concern was not to highlight the cyclical nature of a disease or even the inevitability of crisis but to show that our shock is as predictable as the plagues themselves. The full spectrum of witnessing an epidemic continues to haunt us.

Public health workers might be invisible during an outbreak, doing underpaid but significant work, but even seemingly minor tasks can be essential for relief efforts. At the height of the 2014 Ebola outbreak, the Liberian Ministry of Health initiated a national task force to track and trace the disease, a commission they couldn't afford to carry out alone. After months of tilting with the World Health Organization, the US CDC, and the Action Contre la Faim, contact tracers—some of whom came from the community—assisted these organizations to identify, list, and monitor people potentially exposed to the Ebola virus. These workers took command over their communities to find the sick and dying wherever they could, in the cozy corners of people's homes, where family members provided care, or even in schools, where precocious and determined students learned and proceeded to be with their classmates. The tracers had to be trusted to do the job of keeping Ebola at low incidence and low transmission rates.

While contact tracing has proven effective in some circumstances—particularly for acute and emerging epidemics—the procedure can be limiting when there is an extensive caseload and a lack of staff to carry out contact tracing. But scientists don't necessarily agree on every detail of contact tracing. In an article in *African Jour-*

nal of Health Sciences, Saurabh Shrivastava and Prateek Shrivastava argue that contact tracing was an effective strategy for health workers during the 2014 Ebola outbreak, as it "ultimately aims to reduce the period required to detect and treat a case of an infectious disease and hence significantly minimize the risk of transmission to the subsequent susceptible individuals." Whatever one's stance, the health workers who detected the trajectory of Ebola in Liberian society were due praise. Health workers could intervene and prevent its spread by closely monitoring people suspected of being exposed to the virus. The problem was that there weren't enough healthy medical workers.

Liberia was not alone in the epidemic. When the Ebola outbreak occurred, it was fueled by a chain of structural problems. As the social anthropologist Mike McGovern indicated:

> The human causes begin with the historic underinvestment in the health sector by the three hardest-hit countries and their aid donors. This was compounded by the woeful lack of support provided to national health workers when it became clear that many people were dying from a combination of lack of information and the lack of the most basic medical supplies, like latex gloves.

Though the Liberian government's limitations in the healthcare sector are neither the fault of one person nor of a single institution, the miscarriage of care resulted from a compound of forces, which Adia Benton and Kim Yi Dionne describe thusly: "The legacies of the trans-Atlantic slave trade, colonialism, and structural adjustment programs have helped to shape the political, social and economic environment in which Ebola thrives as much as they shape the response to the crisis itself." The scars of forced displacement and military occupation continue to govern the lives of West Africans, to the point that they fuel collective dispossession.

These legacies are not just abstractions that can be signaled when describing historical injustices; they gesture toward how communities of all sorts are shell-shocked by austerity. The casualty of sickness is not just the battle between a person and the microbe. Still, as Allan Brandt and Martha Gardner have noted, "the relationship between public health and medicine has been characterized by critical tensions, covert hostilities, and, at times, open warfare." Liberia and some surrounding countries have clinical deserts, but more than that, their social conditions and the rapacious extraction within them have given rise to epidemic inequality. The indignity isn't just about death, but that we can come to accept ongoing inequalities. As Ibrahim Abdullah and Ismail Rashid remarked, "Ebola was not simply a deadly viral disease; it was the manifestation of neoliberalism as an affliction, which wreaks havoc in the world's most vulnerable societies." Of course, a disease can be endlessly complicated, but the problem begs us to find a just and layered solution.

Political leaders were reminded that a novel ailment for West Africans could stop health workers from carrying out their jobs. When Ebola arrived during the second wave in December 2014, medical officials were astonished that the disease could not be contained. "Old disease in a new context will bring you surprise," remarked Margaret Chan—then director of the World Health Organization. Several months before, there was a lull in Ebola cases, but when the virus returned, it was unrelenting.

Disease surveillance is the cornerstone of global public health. While it is about optimizing vitality through preventive or acute therapies, public health also reproduces a nation's sovereignty through measures that can appear austere. When the 2014 Ebola outbreak occurred, the catastrophe was not merely contained in the lives lost or the efflux of suffering. Still, some Liberians were dubious about the government's programs. In a survey by Y Care International, 60 percent of respondents said they did not trust health workers or their remedies against Ebola. Some of this was due to a lack of medical supplies—most health centers were closed, and essential businesses and entire neighborhoods were shut during the lockdown. But distrust also emerged as people seeking care were turned away. In one account, a health worker noted that their facility could not accommodate those who relied upon the healthcare system: "Sometimes people feel disappointed and get angry with us when they bring sick people here that we can't treat. They don't even believe us when we say we can't handle some cases." Even when people trusted the system, their trust waned by being turned away.

Politicians and researchers noted that the mandate to avoid contact with individuals who died of Ebola symptoms did not bode well with most people. Death created a significant fissure between health workers and the grieving. At the time, President Sirleaf expressed her unease: "The messages about don't touch the dead, washing your hands, if somebody is sick, leave them—these were all strange things, completely contrary to our tradition and culture." Without a thorough and rich platform in which people could convene, debate, and discuss these issues, it could be challenging to undo medical skepticism. Recognizing the significance of culture and funeral rites, Dr. Muyembe-Tamfum acknowledged that rather than banning burial practices outright, it was significant to acknowledge that medical isolation can be seen as an affront to a population because "by seizing their cadavers we hurt their spirit." He affirmed that safer practices, with gloves and protective material, could provide the spiritual closure that a community needs without sacrificing physical health.

Skepticism is built into the scientific method—people can challenge the authority of research with theory or evidence. At the time, the skepticism and distrust about the Ebola virus were not limited to Liberians alone; the afflicted, who desperately wanted care, could not always access it from the institutions that remained. The late anthropologist Paul Farmer reflected on how the humanitarian aid work system in West Africa and beyond has been broken for years. What became engraved in the minds of the dispossessed and underserved is the withdrawal of a proper livelihood, growing malnutrition, and death from an invisible plague. Even when physicians desperately tried to help, Western media messaging about West African states failed to provide them with the needed aid.

Public health often counters misinformation, but for residents of West Point, the delicate balance between being seen and being respected encapsulated a broader tension within medical relief. Given the novelty of the virus in West Africa and the growing skepticism of the government, some residents were dubious about the disease in the first place. Mistrust wasn't just abstract, but it grew through a mandated quarantine.

The Politics of Quarantine

Quarantine is an act of separation: the imposition to contain or remove someone from society. A quarantine can remove the freedom that a person thought they had, and it can rupture the social foundations that seem commonplace. The word conjures biblical images of lepers, who were seen as pariahs, fussily sequestered in colonies. Quarantine can be a way to exercise power over a probable plague from multiplying. In its most simplified form, quarantine is an attempt to move the feeble or indisposed out of sight. In their book

Until Proven Safe, Geoff Manaugh and Nicola Twilley describe the stakes of quarantine: "While the successful implementation of quarantine can be logistically challenging, the logic behind it remains straightforward: there might be something dangerous inside you—something contagious—on the verge of breaking free." They frame the principle of contagion not solely in terms of its functional qualities but how humans curate their own fortress meant to offset the danger. To protect individuals from ferocious and ungainly microbes metastasizing inside their bodies.

The history of quarantine is not a singular, clear-cut story; instead, it is made up of a broad set of policies that evolved over time. Of course, quarantine has been lauded and criticized. It traces back to the Italian word *quaranta giorni,* "forty days," which was meant to ensure that countries during the time of the bubonic plague kept ships from entering their port after forty days. First enforced in the mid-fourteenth century by Venetians, the period of forty days invoked the Christian rites of sacrifice—such as fasting for forty days—more than any scientific understanding of the life cycle of a microbe. In Venice, sick houses were labeled with wooden crosses in the sixteenth century to make potential visitors aware that the house's residents were infected with the plague. From an epidemiological lens, quarantine can provide physical protection against disease, the simple practice of keeping away from each other, when there is an outbreak. To impose a quarantine isn't effortless; it means containing and possibly banning people and goods from a given space through disciplined restraint. And for centuries, well before people saw the microbe as the genesis of infectious disease, quarantine was used to stave off disease.

Quarantine inflames the artificial borders we create. When Australia and Canada imposed travel restrictions on Guinea, Liberia, and Sierra Leone in October 2014, the WHO asked for their rationale, especially as a global guardian against illness. Since its inception in 1948, the WHO was meant to consolidate efforts to address infectious diseases. A descendant of the International Sanitary Conference—which traced its roots to the mid-nineteenth century—the body was designed to go a step further, to prevent and regulate health globally. Cooperation wasn't necessarily enforced with mandates but given the WHO's principle on established evidenced-based policy of prudence, the travel restrictions for the three West African countries felt hollow. Researchers argued at the time that Global North countries had no reason to carry out travel restrictions. In cases where the Ebola virus patients were treated for the disease in North America and given adequate supportive care, they were less likely to die. Although the cases of patients shipped to US hospitals were significantly fewer than those who remained sick in West Africa, the case-fatality rate observed in the United States was far less than the 60 percent fatality rate observed in West Africa.

As Dr. Timothy Uyeki and his colleagues indicated, "Among the patients with EVD who were cared for in the United States or Europe, close monitoring and aggressive supportive care that included intravenous fluid hydration, correction of electrolyte abnormalities, nutritional support, and critical care management for respiratory and renal failure were needed; 81.5% of these patients who received this care survived." In spite of that, a few North Americans and Europeans who acquired Ebola and were treated outside of West Africa died; media outlets from those countries expressed their concern about Ebola, stoking fear that restricting travel from West Africa was necessary.

In practice, there were other measures to consider besides travel restrictions. According to Dr. Reena Pattani, a physician at the University of Toronto, the fatality boiled down to one thing: "It becomes painfully apparent that the risk of dying from Ebola virus disease is highest where there are systematic failures such as an absent healthcare infrastructure, lack of necessary equipment and a shortage of trained personnel." We can interpret this in several ways, each posing practical and moral questions about quarantine and the challenges countries face in regulating their health concerns. One could focus on the WHO and its role as an international body that was, at the time, making a case for evidence-based Ebola transmission, an infectious disease that spreads through the exchange of fluids, in intimate settings, with an infected person, which is an event that is unlikely to occur during air travel.

Call me a prude, but I rarely exchange blood, saliva, or fecal matter with my fellow air travelers. Perhaps the concern on the part of Canada and Australia, which barred travelers from Guinea, Liberia, and Sierra Leone was not about how humans exchange fluids but the fear that fluids could be exchanged and create a burden on these governments to provide healthcare to their citizens and residents. In other words, they wanted to exert their sovereignty to mitigate the potential problems that might arise when they give treatment to an Ebola patient. While important, these concerns about Ebola would be better addressed if low-income countries had more resources. One might make this case and maintain closed borders; however, global health experience shows that travel bans for non-respiratory infections do very little to contain an infectious illness, especially when a nation lacks a universal healthcare system. Instead, it insulates a nation and conveys the message that the people—not the disease—are dangerous.

The potential harm of the Ebola virus and its violent incursions on the body are horrifying; this is precisely because it pushes the body to a halt. It evokes a condemnation, sometimes of the patient; it "expresses pity but also conveys contempt." Abhorrence isn't easy to swallow—it shapes, on a psychological level, attitudes that determine whether or not someone will be cordoned off. This is precisely what happened to the residents of West Point and the travel restrictions on people who resided in Guinea, Liberia, and Sierra Leone. Not all quarantines are equal; some are imposed as a matter of course—a universal principle applied to all subjects—such as ships entering a fourteenth-century Venetian port, while others aim to cordon off an entire population. For Liberians, the penalty for Ebola was far more significant than a brief moment in time; it was a demonstration of power and a broader disdain for West Africans. That is, how did the two quarantines differ, what can we learn from those differences, how do they showcase different models of state control (national vs. externally imposed)?

For West Africans, the 2014 Ebola crisis demonstrated that Liberian leaders valued security over care. The travel restrictions were not just about exclusion, nor did they unravel new hierarchies of governance. It is worth asking if it was necessary to isolate Liberians to contain the Ebola virus. Did this policy save lives? Chief among these questions is how humanitarian bodies use military troops during a public health crisis. The armament of Liberia during the initial stages of the 2014 Ebola outbreak was lofty, a heavy intrusion into the lives of everyday individuals—but also an indication that they have—on an allegorical and corporal level—disease and that they could be less contagious if they were contained. In her book *Pandemic,* the journalist Sonia Shah remarked about cholera, "humans can carry the microbe, of course, but only so far." What she points out is that microorganisms continue to colonize our bodies, hoping to replicate and mutate within the body, producing new progeny and ensuring the survival of their kin. But their viability has a threshold— they cannot travel or remain viable too long without a host.

During the peak of the lockdown in Liberia, public health officials had to think about how the clashes between civilians and the military—not just as a social collision but the ways that the country's recent history surfaced and cast doubt on the protocols. A state of emergency, the legal decision to regulate and curtail people, can be viewed as collective punishment, especially as people's lives are stalled and disciplined. When I spoke with Tolbert Nyenswah, former Deputy Minister of Health, he recalled that when the administration adopted military intervention by implementing quarantine and curfew, it triggered brute force of the past given that "the memory [of the Civil War] was still fresh." In the frenzy of trying to contain the virus, the commitment to initiate these policies had more to do with the fact that hospitals and clinics could not accommodate every patient who contracted Ebola.

The looming dystopia is not just about Ebola itself; instead, it speaks to the climate of African containment, the military occupation, and when people are forced to fend for themselves. The details were stark for Liberians who were met with aggression—their freedom wasn't theirs to make. Instead, global panic meant that they could be subjected to neglect. They weren't unparalleled. As scholar Dr. Adia Benton noted, the military logic of the Ebola intervention in Sierra Leone resulted in a war between the virus and the people, one where the military system was built up while public health institutions were neglected. This militaristic blueprint, according to Benton, was austerity and neocolonialism in action. On September 9, 2014, President Sirleaf petitioned the US army to intervene in Liberia. Following this request, President Barack Obama conceded and sent three thousand troops to Liberia, a task initially meant to erect new Ebola treatment units and help with transportation and other logistics. In reality, military presence along the Liberian border and neighborhoods like West Point conjured the ghosts of the Civil War.

As Ebola deaths swelled, the disease was subject to a conventional outbreak narrative: its ascent and spread, scientists trying to trace the geographic dispersal of infection, identifying a disease carrier, and a massive public health campaign that ended with its containment. After cases were detected in Europe and North America, Western governments suspended flights from West Africa. They mounted an international effort to restrict residents of several West African countries from being able to travel to Global North countries. With no vaccine or cure at the time, Ebola took up considerable attention, which put stress on the healthcare system for managing other diseases. The Liberian president decided that the only way to prevent the disease from spreading was to contain it.

The epidemic was draining, but not for the reasons North American countries expected. The acute epidemic revealed severe fault lines in the theory and practice of global public health: National governments experienced a lack of solidarity even when few people died under care. This was the conundrum that Liberians had to deal with—the perception that they were dangerous—or rather, contagious, which meant that they had to live in a state of exigency.

The lockdown of West Point and travel restrictions on three West African countries were not a single event. Still, they revealed many aspects of Liberia's history—the ghosts of US influence, a starved public health system, and the memory of the Civil War—that collectively demonstrate who would be contained and who could leave the country. The Ebola outbreak was not simply about one African country but about the multifaceted dimensions of who is allowed to travel, who is protected, and who is provided agency. Despite these efforts, the Ebola spread like a wave, moving through hospitals, and leaving some people prostrate.

People perish every day; this is part of what it means to inhabit a world where microbes outnumber humans. Whether or not someone survived the Ebola virus in 2014 depended heavily on whether they got essential treatment, including clean water. Travel restrictions were not the only source of harm. Rather, typecasting meant that Liberians were misrepresented or sensationalized.

The Obsession with Monkey Meat

When I was researching Ebola on the internet, I found that some journalists are obsessed with monkeys—especially when concerning Africa or Africans. The Western accounts played on EuroAmerican audiences' ignorance and relied on shock value to get the viewer to keep watching. One video starts in Monrovia, Liberia, with the lyrics, "Ebola in Town; don't touch your friend," blaring in the background. But it begins with a question. A male voice, off camera, speaks to the men and asks: "Do you guys ever worry about Ebola?"

The two men, wearing short-sleeved shirts, chewing, respond simultaneously, "No."

Several other people remark, "No Ebola in Liberia."

The program was produced by VICE Media in 2014. Kaj Larsen, the journalist and narrator, had one concern: to understand if and how monkey meat contributed to the spread of the Ebola virus. To an extent, the query is a legitimate question and is worth pursuing. But how it was initiated can be seen as morally dubious.

In July 2014, the Liberian government banned the hunting and sale of bushmeat, fearing that Ebola would spread through the sale and consumption of the commodity. This meant that anyone caught with the provisions could be imprisoned. Larsen, being aware that the Forestry Department was tasked to search people's property and seize bushmeat if they saw it, asked several scientists based in Liberia if this was valid. Sitting with Dr. Joseph Fair, a virologist at the Liberian Institute for Biomedical Research, Larsen asked if living alongside non-domesticated animals contributed to the spread of the virus. Dr. Fair pointed out that humans have co-evolved with animals throughout their existence and while the Ebola virus could spread through meat, the main transmission was from other infected humans.

In one study published in *Social Science & Medicine,* researchers surmised that banning bushmeat exacerbated growing distrust in Sierra Leone toward people living in low-income communities and the countryside. The authors argued that despite the legacy of meat posing a risk for the spread of zoonotic disease, the bushmeat ban "contradicted the experiences of targeted publics, who consumed wild meat without incident." In the midst of the outbreak and well after, these scientific insights seem to pierce through the heart of Larsen's query—the relationship between monkey meat and Ebola. However, this expertise was not enough. Sensationalization may have been more entertaining for the journalist.

Returning to the video, Larsen seems dubious of the people who eat bushmeat, though he tries to get the perspective from Liberians directly. In one conversation, Imanoel Nagbe, a charismatic journalist with a bright smile and wearing a purple shirt, notes that "the feeling in Liberia is that Ebola never came here." With a wry smile, he asserts that the Ebola virus warning is a scam, an attempt by the government to get money. As the camera rolls, Nagbe takes Larsen to a bustling market, where women put their produce and meat on display. A man in a red shirt with a bullhorn and tambourine walks up and down the aisles. Larsen and Nagbe walk through the aisle of the bazaar with the camera, homing in on a group of Liberian women selling meat.

Nagbe and Larsen approach the women, and Nagbe asks, "Has the meat caused you any problems?"

One of the women responds, "No."

Another market woman leaves her section, and enters the conversation, denies the meat is an issue, and affirms, "My children and myself have been eating it [bushmeat]."

The people in the bazaar are confident in saying they do not have Ebola. They look relatively healthy and exert heaps of energy. They are neither embarrassed nor reticent when asked about their health status. In another encounter, Larsen accompanies a group of men in central Monrovia; they openly admit that they love bushmeat and decide to consume some of it on camera. One man holds the carcass, and a bit of the liquid is expelled from the meat and shoots into the air. Larsen looks disgusted, croaks, and remarks that "I'm getting pegged with Ebola." Larsen then asks him whether he thinks Ebola is real. The men quickly respond that they don't. They acknowledge that the government has indicated that Ebola victims have a hemorrhagic fever but given that they have been eating the meat without acquiring the fever, they doubt that the disease exists. Their view reflects the rupture between public health messaging and their everyday experience.

The video ends with a premonition by Larsen: "Some human, somewhere, likely here in Africa, is going to eat a monkey with a disease that has never been seen before. Super Ebola, a new strain of AIDS, and according to [Dr.] Joseph [Fair], that virus could kill a third of the population. It's not the disease we know; it's the disease we don't."

In my mind, Larsen's warning is full of indignation, not only over the meat that people consume but at the people themselves. Their skepticism about Ebola fuels his horror. What we see is that a group of people, who on the outside appear healthy, are eating a snack on camera. Although the few people appearing on-screen were allowed to express their perspectives, it seems to me the journalist denigrated them. This VICE video might seem crass and exceptional, but they can be seen as a metonym for how Liberians and Ebola are represented and sensationalized. Reporters seem to have used a predictable story to dismiss West Africans, to zero in on their discomfort surrounding some people's denial of the outbreak.

North American disdain for Africans appears to be at the heart of these images, channeled through such fantastical claims about Liberians and their meat. This gestures to William Ian Miller's conceptualization of "disgust"—whereby people allege or identify boundaries about one's life and express moral boundaries around it. Disgust is an emotion we all possess; it shapes the relationships we form, our behavior, and our desire. However, the power to name and put forward claims and concerns about a group that may dehumanize them seems to speak to the old-fashioned conceit that African practices will spread infection.

Visual culture in public health discourse is not just a matter of shaping policy but perception. Viewed generously, the VICE video satiates the viewer's curiosity: What some members of the Liberian population think, as presented by Larsen, is presented with little subtlety or refinement. At the same time, the visual grammar bristles against an uncomfortable tension between misinformation and lived experience—and who has the power to frame that narrative. In Susan Sontag's book *On Photography,* the author provides an indiscriminate but intellectually rigorous take on images and their signature ability to establish meaning in different cultures. "To photograph is to appropriate the thing photographed," Sontag writes. "It means putting oneself into a certain relation to the world that feels like knowledge—and therefore, like power." It seems to me Larsen, in effect, puts himself above the people, without unpacking why they believed what they did or how their distrust emerged.

Precisely what did the VICE documentary on Ebola accomplish? Was it an appeal to provide better public healthcare, or was it meant to confirm the biases people already have? Did it help inform Liberians about the medical context or was it a way of fueling American fears that bushmeat was the sole source of the Ebola crisis? Racial overtones slice through the obsession over bushmeat; these practices can be seen as a stand-in for outright racism. There is an obsession in demonstrating that bushmeat is the problem, even if US Americans also eat it, only they call it game.

Cartoons produced during the 2014 Ebola outbreak ridiculed containment. In one satirical piece for the *New York Times*, Patrick Chappatte drew a map of Africa, bare but full of meaning. The continent's borders are made with rescue tents; inside, there are two ill people next to an Ebola sign, two healthcare workers in medical gear carrying a corpse, and others watching people positioned outside the tent. Those outside exclaim: "You're under quarantine . . . for your own good." Presumably, the West aiming to contain the virus. The message, while gently mocked here, is clear: Africans have to remain inside their walls.

Dealing with Ebola. Patrick Chappatte, *New York Times*

Chappatte's commentary, and others like his, force us to grapple with the assumptions we make about mobility. The infected are mostly positioned in West Africa, with the dead being carried away by two health workers in hazmat suits and two people sitting on the ground near an Ebola sign. One person lies unconscious on the floor and a woman and child stand absentmindedly near the health-care workers carrying the corpse. The virus is residing inside the boundaries of Africa—constructed with a brown tent—while two unaffected observers remain outside the continent, letting those suffering the impacts of the virus know they are under quarantine. The Africans, here, remain inside the boundaries of the continent, and non-Africans, self-imposed experts and saviors, are instructing them about public health measures to which they would not subject themselves.

The cartoonist also casts an unflattering light on multiple corners of government and international response. Fear, not science, drives the decisions made at population and leadership levels, much as it takes place so often in the real world. As a result, responses are often as uneven as they are unprecedented: entire nations condemned by travel restrictions, health staff left to die, untested medications used (or not used) in controversial ways, humanitarian workers forced to isolate upon their return home, and military force employed to sometimes disturbing ends.

Return of the Lockdown

"There was no proper sanitation and hygiene, water, and a lot of those things available," remarked Tolbert Nyenswah. "So, we had to reactivate our whole water, sanitation, and hygiene team supporting these facilities. And I carry on IPC infection prevention and control and giving healthcare workers the personal protective gear, gloves, soap sanitizers, chlorine, everything." When I spoke with Mr. Nyenswah, he was assertive, expressing deep confidence about emergency preparedness during the 2014 outbreak. In 2015, he became Deputy Minister of Health; during that period, he helped establish the National Public Health Institute. By 2017, he was appointed the Director General of the Institute. During the 2014 Ebola outbreak, an emergency team of clinicians and public health professionals worked on diagnostics and supply chains—anything to ease the burden on countries that may have had trouble mounting a robust response. But in their earliest statements, microbes contained multitudes, to paraphrase the science writer Ed Yong.

When President Sirleaf established an investment plan for building a resilient healthcare system, this meant constructing, according to Mr. Nyenswah, a durable health surveillance system, expanding everyday healthcare facilities, and increasing the number of physicians in Liberia. Although primary health services such as clinics are free, there is no universal healthcare in Liberia. Even in government-run facilities, people can end up paying up to 40 percent of costs.

Nearly everything Mr. Nyenswah told me reflected on the lockdown, the dynamic forces of public health, and the tension between quarantine and mobility. The early days of the outbreak evinced a broader predicament within aid work: As public health officials observed, implementing a neighborhood lockdown gave an outbreak little room to recede. And so, their decisions yielded a tide of protests and two responses that compromised the health and safety of everyone.

For Mr. Nyenswah, "[the lockdown] was a mistake. And the government admitted to doing that. It was a mistake. I think the best strategy was the community-based initiative aimed at people." He elaborated that the Ministry of Health, well after the outbreak, recognized this. Rather than continue lockdown measures, they expanded community-based programs, sometimes compensating people for wages lost when hospitalized. The question remains, how do you engage with a community that does not trust public health officials? Mr. Nyenswah spoke with concern, dwelling on what it would mean to address the inadequacies in the healthcare system and move from a patient-centered approach to community-based medicine, which centers the most marginalized.

On May 2015, the WHO declared Liberia Ebola-free. But this announcement was rescinded multiple times as cases re-emerged. By the outbreak's end, Guinea, Liberia, and Sierra Leone were the three most affected countries. One year later, in Sierra Leone, the government established a *cordon sanitaire* in several districts, with officials hoping that non-Ebola healthcare, goods, and services would remain infection-free. The outbreak lasted two and a half years, resulting in nearly 30,000 confirmed reported cases and over 11,000 deaths. The immediate effect was particularly harrowing for Liberia. Of the West African countries, Liberia had the most reported deaths, with nearly 5,000 deceased. Though some aid workers estimate that the numbers are higher. The distinction between life and death during that period had more to do with inequities in health systems than with the disease itself. There were approximately thirty cases and fifteen deaths outside of West Africa.

The epidemic revealed severe fault lines at all levels of public health: National governments were shown to be vulnerable and unprepared for calamity at this scale, the World Health Organization was roundly condemned for its ineffectiveness. At the same time, Liberia created the National Public Health Institute, and officials on the African continent established the African CDC in the aftermath of the Ebola epidemic. When the 2014 outbreak occurred, commentators noted that the lack of treatment for Ebola was disturbing and reflected the calculus of drug development companies for treating diseases. Finance writer James Surowiecki noted:

> That means that they have an incentive to target diseases that affect wealthier people (above all, people in the developed world), who can afford to pay a lot. They have an incentive to make drugs that many people will take. And they have an incentive to make drugs that people will take regularly for a long time—drugs like statins.

Surowiecki's observation is not entirely new, nor does it complicate the view of pharmaceutical companies' conflicting priorities, which are driven by profit over goodwill. Nevertheless, vaccines offer a shell of protection and can save lives. Do most European or American scientists have the same enthusiasm to save African lives as they do their own? Probably not, but what happens when they try? As of January 2023, two licensed Ebola vaccines are effective against the Zaire ebolavirus species—Ervebo and Zabdeno. While there is an immense benefit in vaccine development, examining how universal quality healthcare can be provided to everyone would be beneficial beyond simply disease prevention.

There were limits to the lockdown in West Point, especially given that the measure did not initially come with new clinics, hospitals, or even training for local staff. It showed that there was no single remedy for the virus and that the more significant issue had to do with the re-emergence of the disease. Harvard University historian Dr. Emmanuel Akyeampong noted, "Most of the diseases that were present in 1900 have resurfaced, except smallpox. Major environmental changes, such as global warming and changes in land use, have transformed higher and cooler areas in Africa that were previously malaria-free into malaria zones." Dr. Akyeampong confessed something that continues to ring true—so long as the environment changes for the worse, we will live side by side with virulent plagues. It is not the fault of the meats we eat but of an Earth that humans have made uninhabitable.

Like many people trained in public health, I have had to meditate on the limits of charity and the role of non-governmental work in the Global South—which has become commonplace in countries like Liberia. While emergency interventions during an epidemic can provide relief, there are limits to these programs because of their inability to provide a full range of services. At the core of these issues is that they fail to disentangle the structural issues that make people ill. Perhaps we need to think beyond bans and imagine what it would mean to provide universal healthcare to everyone globally.

In his 1938 essay, "The World's One Hope," the German playwright Bertolt Brecht remarked, "All those who have thought about the bad state of things refuse to appeal to the compassion of one group of people for another. But the compassion of the oppressed for the oppressed is indispensable. It is the world's one hope." That compassion was missing by some Western governments and media when West African countries were banned.

The 2014 Ebola outbreak in Liberia is a story of a country that overcame an epidemic. The residents of West Point battled with a nascent disease in the context of an underfunded public health system—and then had to grapple with all the ways that global authorities sought to contain the disease and further pushed Africans to the periphery. Nonetheless, the state ultimately recognized that the quarantine failed. Public health should be a buoy riding the harsh sea waves, tending to the sick and exhausted. Ultimately, some health workers who cared for Ebola patients have told their stories, expressed disdain, and urged for more resources.

To point this out is to question how and why these people were embedded in the lockdown.

A year after his recovery from Ebola, Dr. Philip Ireland reflected: "Health workers were largely underpaid and poorly prepared." The disease made its way through the hospital and an underfunded health system, killing more than 150 healthcare workers in Liberia. Dr. Ireland was closely attuned to the fine balance between life and death during an epidemic and was reminded that Liberia needs "to address the health problems that have been neglected under the onslaught" of the Ebola epidemic.

Epidemics, even the familiar ones, force healthcare workers to reflect on matters of care.

Chapter 6

RELENTLESS

Mask no difficulties, mistakes, failures. Claim no easy Victories
. . . Our experience has shown us that in the general framework of daily struggle this battle against ourselves, this struggle
against our own weaknesses . . . is the most difficult of all.

—Amílcar Cabral

F OR FELLOW MILLENNIALS LIKE MYSELF, 2020 was crip-
pling, yet another lumbering dilemma that thwarted the possibility of wealth or stability for our generation, akin to the
global financial crash of 2008. Knowing that I wasn't alone, I began
reaching out to people on the peripheries of my network—and far
outside of it—to inhabitants in Germany and figuring out how they
changed. More than anything, I wanted to know about the most
vulnerable people in Germany. What did it mean for individuals who
were already socially isolated for other reasons to also find love and
resilience during the pandemic? I wanted to find the thread that
connected us and made us whole. So, for a year, I regularly contacted
several individuals living in Germany—speaking with them about
the lockdown, their lives, and what they shouldered.

"Grief, then, for these deceased others might align some of us,
for the first time, with the living," wrote Claudia Rankine in her
2015 essay "The Condition of Black Life Is One of Mourning." The
essay's core was to discuss the historical and current premature
death of African Americans. Re-reading the essay in 2020, I absorbed

the grief that Rankine laid out, and slowly filled with rage. Mostly because I saw how anti-Blackness and ruthlessness toward African-descended people was not just a Black American phenomenon. It was visible for West Africans who were targeted by European mercenaries, Black people who crossed the Mediterranean Sea by boat, and Afro-Europeans who lived in Germany. For African-descended folks who lived in the country that did not always recognize their full humanity. I wanted to understand how people, who were vulnerable in some way, were more captive.

These are three excerpts from several Black people who live in different parts of Germany—a sex worker, an asylum seeker, and a cancer survivor. For the three women I spoke to, their stories felt familiar. Covid-19 had augmented their social confinement. The relative isolation they felt during the pandemic was an act of witnessing, in real-time, growing economic uncertainty, illness, and grief. They carried deep wounds seeping into their necks, backs, and hearts. It was a pain that reverberated whenever they had to scramble to pay rent, when they had to advocate for their health, or when they were fearful of contagion. My conversations with Rose, Sandra, and Ava had come to an end, and I wanted to read through their lives, to see how their stories were rendered within the ongoing pandemic.

In the first year and a half of the Covid-19 pandemic, our conversations were an intimation about Black life in Germany, in its fullness and complexity. They reflected the power, privilege, and injustice that are present and absent for individuals whose identities are apparent through their idiosyncrasies. These discussions resonated because they went beyond the formless news reports of the first two years of the pandemic. Instead, the people spoke about their confinement in ways that showed their beauty and grace. Some of our chats were extended and others were brief. Sometimes we felt broken on both ends, but through these moments of trust, even if our lives felt unsteady, we could sustain a dialogue during the pandemic. Writing them into a broader story of confinement wasn't enough. Their lives were a constant negotiation between the real pain they felt at the time, and this constant need to find joy as Black people living in Germany. This was a period where I had to listen and wind through the intimate lives of strangers who made an immense effort to thrive.

Sex Work Is Work

Rose (this is a pseudonym), a sex worker in her mid-forties from the United Kingdom, had worked for years in the German sex industry, from community workshops that centered queer people to one-on-one sexual services. Receding into her home during the Covid-19 pandemic shifted her intimate and professional life, but it also had her reflect on whom she no longer had access to—the country she came from. Born in East London and raised in a predominantly white English suburb, Rose lamented her inability to visit her family during the pandemic. Hailing from the Caribbean, her parents were the first of their family to settle in the UK. When her grandmother directed her father to leave the Caribbean she said, "he came kicking and screaming ... tried to survive [the UK] after being a really small island boy that wanted to stay on this small island." Her parents studied and settled in England like many Afro-Caribbean people, helping to augment the country that had colonized theirs.

Rose's childhood wasn't always happy, but she felt she had many opportunities growing up in the UK. While studying to be a chef, she caught a traveling bug—spending time in France and Germany. She cited racism and misogyny as difficult encounters that she had to cope with during her first visit to the country, and I asked her why she resided in a country where she was targeted and deeply hurt by discrimination. This wasn't a major issue because of what Germany offered:

> I wanted to be everything that my world at the time was not as a child, it was not diverse, it was not well integrated, and it was not inclusive of me. And I wanted to find somewhere that was that. And I found out it is you who create your world, you create your universe. In a place you want to feel comfortable, surround yourself with like-minded folks.

There are many places in Berlin where people can revel in their kink—whether it is full frontal exhibitionism or queer play parties—but there are few places in the city where people can safely do so, and even fewer for people of color who want to experiment with their sexuality. Rose knew the space she envisioned had to be imaginative, a blend of consent and pleasure, a haven for sex-positive people who did not just want to be in the margins but at the center of kink parties.

The British philosopher Iris Murdoch never had any reservations about discussing fornication. In her novel *The Nice and the Good*, she affirmed, "sex comes to most of us with a twist." The first time I saw Rose in her kink studio in Berlin, embossed with confidence, I saw how she curated both a safe and sensual practice. This was in January 2020, when Rose invited me and my interlocutor—a filmmaker—to document her work for a short documentary we were working on.

It was a chilly winter afternoon, blunted by the grayish skies and burly winds. When we entered the vast studio, I stripped the space for its purpose. The brick-red walls were decorated with colorful fabrics, and the floor was strewn with ropes and harnesses that more resembled a fitness hound's CrossFit workout than the prurient activities they were meant for. My collaborator and I needed to know: How did her kink practice bring her joy?

Was it a feeling that flowed freely through her soul; was it a sensation that was sentimental or ecumenical?

Rather than tell us, she showed us. With the help of another person, she showed us how the kink studio functioned, as a center for expression and release regardless of background.

For ten minutes, using the rope, they reveled in light touch, engaged in intimate bondage, sat on the floor and embraced with their eyes closed. With each passing moment, their smiles grew a little wider, a grin that ran toward euphoria. This space was an opportunity for people to work creatively with their bodies, in whatever form they took. Salient from that was the power of pleasure-driven commitment, brought forth by a service Rose as an entrepreneur and administrator could provide. This was two months before Germany declared a country-wide lockdown, and at the time, Rose still felt grateful to offer some respite.

People like Rose echoed the virtues of Black feminist thought. In her seminal essay "The Uses of the Erotic," Audre Lorde makes the case for unfolding the eros: "Recognizing the power of the erotic within our lives can give us the energy to pursue genuine change within our world, rather than merely settling for a shift in characters in the same dreary drama." The erotic had been a core feature of Rose's life; when Covid-19 cases occurred in Germany, that world would come to a halt.

By October 2020, Rose reported that she was falling straight back into survival mode, which often led to replicating trauma-related patterns. Having just lost a friend to death by suicide, she was trying to find some joy, to reconnect to the physical spaces that revived her spirits. "What can I do for myself now? What can I revel in now?" she asked me. She used this time to do online singing lessons and attend an online workshop on Japanese bondage centered on queer femmes. For the most part, these activities were her trying to navigate what she described as an "emotional curatorial process."

Seven months into the pandemic, the rapture that she expressed when I saw her that January afternoon was nearly gone. Her brow furrowed as we spoke online, and she reminded me that there was little room for the collective community that dared to have the freedom to translate an intimate touch. Given the lockdown, Rose turned to the internet like many kink-positive people. Though Rose is aware of websites such as OnlyFans and Backpage, she and her community gathered in other forums, on platforms trying to decolonize desire and promote anti-racism in the kink community. These were virtual spaces that expanded the boundaries of lust. For many sex workers who could not perform their physical labor in person, turning to the digital world to share tips and reflect on the industry brought them to life. Although Rose and her kink interlocutors were able to organize sex-positive workshops, having had less work during the initial months of the pandemic had significantly diminished her income. These online spaces meant that Rose could conduct workshops, which was initially challenging, but over time, as she got better, she noticed that she was adapting to the lockdown. "I've had the opportunity to make more conscious choices of what I do and don't do. Online, I'm open to doing the work."

In Germany, she could insulate herself from the health disparities commonly inflicted upon people "like herself—Black, Asian, and other minority people—in the United Kingdom. But the data was not the only part that was upsetting; it was the understanding that Covid precautions under then–prime minister Boris Johnson were callously conceived and administered, protective of our most privileged at the cost of our least. As the *Guardian* health editor Andrew Gregory reported in 2022, "the absence of a specific programme of work investigating how racial inequality and racism affected millions of Britons over the last two years has caused anger and prompted half a dozen community leaders to raise their concerns." Given this, Rose did not want to risk the chance of getting Covid—either from clients or lovers. "I miss being able to see my friends, and not being able to keep the connections that I want from a physical perspective is having a real detrimental effect on me."

In March 2020, Germany banned large gatherings, closed schools, and prohibited prostitution as measures against the spread of Covid-19. As it was for all of the country's estimated half million registered sex workers, Rose's labor was inextricably changed by the virus. Prior to this time Germany had enacted policies to establish legal measures to make sex work safer for those consenting to engage. In 2001, the German government passed the Prostitute Act (*Prostitutionsgesetz*), an affirmation that exchanging money for sexual acts would not only be decriminalized, but protected. Sex workers could make claims to the state if they were not compensated properly or if they were exploited in their workplace. And for the first time, they could freely disclose their profession as they sought access to public healthcare, unemployment benefits, and a pension. By 2017, the Prostitution Protection Act was put into place and had grown to encompass a set of compulsory regulations—to access the protection granted, sex workers would need to register with the state, undergo compulsory medical evaluations, and seek yearly renewal of licenses to practice their labor.

The experience of sex workers in Germany contrasts significantly with the vast majority of countries around the world, including the United States, where sex work is illegal or severely limited (in some counties in Nevada, prostitution is legal). For Rose, being a sex worker in Germany wasn't just about the legality or relative convenience of the labor—compared to the UK—it was about the ability to access the community and support of a professionalized industry. She was empowered by the humanity of the people who lived the same experiences she did. She reveled in a space that fostered inclusion for queer people, and women, and immigrants, and ethnic minorities. But curating that community was not always easy.

"Sexual expression is involved in care. It recognizes that empathy is important," Rose told me. The pandemic taught her to think about the messiness of society. But more than anything, it brought renewed awareness to the struggle for recognition she and her compatriots had been wrestling with all along.

Rose never shies away from these material realities of work if she does not reflect on how the Covid-19 pandemic has been interpreted differently by the people in her life. She understood that her ability to transition to an online platform was also a matter of survival. Rose indicated that "some people don't have as much of a choice, like no money, not enough support. Like you're going to go do what you need to do to be able to pay rent and put food on your table." During the winter of 2021, when there was a Covid scare in her apartment, everyone had to isolate. Rose tried to be mindful and careful, but she grew frustrated when one of her roommates disregarded the isolation protocols that the household agreed on and dismissed Covid-19 protocols stating that his partner had tested positive for the novel coronavirus and "she was fine."

By late spring 2021, Rose was distraught. "I am somebody who doesn't do well in confined spaces and that's not just physical space; I mean mental, emotional, spiritual space." But it was also difficult not to see how "the world over has an undermining and degrading view of what sex is in general." When I asked her if she would ever move back to the UK, she swiftly responded, "No." Indicating that "the way the UK looks now and the way the UK seems to be heading is not a place I wanna go." But the issue wasn't just about her, it was about how she saw her family living. She noted,

> I watch my brothers, all three of them who are all smart young men who struggle to get jobs, and who struggle to live. They're just existing. And that's what I walked away from. I don't want to exist. I want to live. I want to be out to nurture myself, take care of myself, and have more freedom. And that's surprising in Berlin. I have more access to things in Germany than I would if I were in the UK.

By the fall of 2021, Germany had sped up. Restaurants were open; patrons were drinking pilsners from their neighborhood bars; nightclubs were again animated from Friday to Monday morning. As the city became spirited, people in the sex industry also got back to work. Rose was enlivened and wanted to politicize her work further, but also recognized that more needed to be done. She believed that sex work needs to be legal everywhere but more specifically for people "to acknowledge its value and its worth of knowing your body." But seeing things open up, some were also committed to a broader message and Rose pointed out that:

> Sex work is work, and there's joy in that work. People can play, create a role and fantasies, and play them out. Nobody seems to acknowledge the fact that sex workers are part physical embodiment, part therapy, part creation, and part desire.

In spite of the pandemic, some people were ravenous for sex. They yearned for sensuous encounters outside the generic virtual courtship: a meeting of eyes across a subway platform or a slight brush on the arm. Some people wanted to feel the sweat of a potential lover slide onto their flesh, tied to a bed stand, their toes sucked, or even cuddle with a bonded intimacy that is abundant, expansive, and perennial.

Rose was finding again and again that the containment of epidemics pathologizes non-white people, who are often perceived as the source of contagion, even as, through internment by the majority, they are made more susceptible to infection. Sex workers make these experiences possible; they curate kinks—virtual and physical. They help people manifest fear of sexual reverie; they help deepen people's capacity to engage in a newfound candor.

Finding Refuge

My first visit to Sandra's (a pseudonym) *Flüchtlingslager*, a refugee camp, was in August 2021, in the outskirts of southern Brandenburg. It was a mild Berlin summer—neither sweltering nor humid. That day, the sky had a grayish haze. As I journeyed by train from Berlin to rural Brandenburg, the province where the capital city resides, I ushered myself into the monotony of the German countryside. The landscape was unimpressive, a lowland interspersed with marshes, farms, and hamlets.

Given the fact that Germany had a strict Covid-19 lockdown— limited public gatherings—it was difficult to visit Sandra for the first half of 2021. This was the first time that I could make the journey; my previous attempts to visit were postponed because of the national railway strikes. "Overwhelmed with being overworked and underpaid, the regional train workers had been demanding better pay. Like most essential workers, they echoed the solemnity that surrounded us: They wanted recompense for the unpleasant task of working harder for less. For asylum seekers like Sandra, living in a refugee camp in rural Germany, when the country went into a lockdown, the world that they knew had shrunk. She was pulled into a solitary undertow, hoping that her case would be approved and that she would be able to establish a permanent home in Germany. The Covid-19 lockdown decelerated many asylum cases because government workers pegged away at reduced hours many away from the office. This meant that asylum seekers awaiting their cases might have longer waiting times than expected to receive a decision about their legal case.

When I first moved to Germany, I believed their migration policy was humane. Compared to the United States, where I was born, and the UK, where my partner was born, Germany appeared pro-refugee by accepting large numbers in 2015. In 2022, Germany had 3.3 million refugees and people living in exile; altogether, there were 11.8 million foreigners living in Germany. Since 2017, I have been one of them. But I knew that my migration was significantly different from the many asylum seekers and immigrants from the Global South that I befriended. I knew that the lockdown was fundamentally different for me (in form), but I wanted to understand how other migrants fared.

The salmon-colored train station was as pitiful as the desolate surroundings. The words *meth schlampe* (meth slut) were spray-painted on one wall. When I stepped off the train, I saw the large clock resting above the main station doors. Next to the platform, there was a capacious sign on the upper left side that read Bahnhofs gebäude zu verkaufen (station building for sale). To the lower right was a sign that read Historischer Lehrstand (historical trail), indicating that the town lay on a historic route between Dresden and Berlin. As I strolled through the streets, I began to see political ads for the upcoming German parliamentary elections. Silvio Wolf, one of the candidates for Alternative für Deutschland, the largest far-right party to gain major political power since the Second World War, marked his message. "Gleicher Lohn für Gleiche Arbeit; Ost/West" (Equal pay for equal work; East/West). He was a toolmaker who prided himself on being "down to earth." When I looked up his campaign message on the AfD website, he claimed: "The old parties have been telling citizens for years that AfD members and politicians are 'Nazis,' and act against the Basic Law and are not open to the world. Should the citizens themselves question whether their homeland has changed for the better or worse due to the politics of the old parties?" It wasn't clear which policies he meant, but in 2013, when the AfD first formed, they wanted Germany to accept fewer refugees and to deport migrants who were charged with criminal offenses. Their main objective was to stop migration.

The silent undertones of the town were suffocating.

As I continued walking toward Sandra's *Lager*, I saw an open field freshly cultivated with wheat. There were no tractors or signs of other people, and I noticed I was the only pedestrian. Walking down the path, I pondered what it would be like to walk here during the middle of the winter when the town gets dark at 3 p.m. *Did Sandra feel safe walking here?* I strolled past a strip mall. Just like in American, the goth kids were smoking cigarettes in the parking lot.

In 2020, Germany had two lockdowns, which meant the borders were closed for the first one and heavily restricted—like other EU countries—for foreign visitors. Refugees were some of the most vulnerable people during the first lockdown. With people sharing bathrooms and kitchens, social distancing proved impossible. The public health measures residents were supposed to adhere to could easily fall apart.

Sandra was running late when I reached the *Lager*. Her German class—which she took through the *Volkshochschule*, the community college—had run overtime. From the entrance, the building and the entrance were openly exposed. I caught the sight of a seven-year-old girl playing with a jump rope, her ponytail bouncing. She was near two women wearing headscarves who sat underneath the basketball court. Sitting next to a modest outdoor table, the two women were chatting and enjoying the light summer breeze.

When Sandra arrived, her face looked bright. The last time we saw each other was the winter before in Berlin, right before her surgery. This time, her hair was braided, and she was carrying her school items: a backpack, water bottle, and mobile phone. "I'm so excited to be learning German finally," she remarked. I was taken aback. "Why weren't you able to do this before?" I inquired. "My social worker had to approve the process before since it is not clear my asylum case would be approved. But since it looks hopeful, I was granted permission to take German classes," Sandra responded. As we moved toward the entrance of the building, I saw a sign that read **Schützen wir uns** (Let's protect ourselves). As I scanned the other signs, the texts were entirely in German. None were in Arabic, Farsi, Pashto, or English—the languages that most of the other residents spoke. Sandra admitted that she was finally able to understand some of the signs.

"How long are you visiting for?" the guard inquired. "Only for a couple of hours," I responded. "Give me your passport," he replied. I handed him my German residence card instead. After signing the visitor's log, we walked through the corridor. The fluorescent lights flickered a bit. Some doors had a small carpet and shoes in front of them. We passed by the communal kitchen, two ovens, and the communal bathroom. When we arrived at Sandra's room, she turned on the lights. Here, she kept her most precious items: a vivacious vine plant, two chairs, a large carpet, and a comfortable sofa. After corresponding for a year, I finally had a chance to see where Sandra spent the majority of the Covid lockdown.

Refugees and asylum seekers globally have to live in conditions that make them vulnerable to communicable diseases due to cramped living quarters and precarious working conditions. With limited access to healthcare and a lack of confidence in government authorities, many tried to navigate the residential facilities that were not designed for a prolonged lockdown. In a 2021 retrospective study of refugees living in Greece, researchers found that the risk of Covid-19 infection was significantly higher for asylum seekers due to overcrowded housing. Even when people wanted to practice social distancing measures, the sanitary measures were brutal to put into practice.

Born and raised in Central Kenya in the mid-1970s, Sandra migrated to Germany in June 2019. When she fled Kenya, Sandra told me, "I left running for my life." Migrating had been a choice she needed to make, given that she no longer felt safe in Kenya, even though that meant that she would leave her two daughters behind. The chain of events that led to her exodus—her family quandary, being unmarried, and her father's death—complicated her status within her ethnic community. Her decision to remain unmarried with two children stirred outrage in her village.

Sandra spent the first month in Berlin; she was constantly under the knife. When she first arrived, she was examined for tuberculosis, with the assumption that people coming from the African continent are more likely to have this disease. The X-ray was clear: Sandra did not have tuberculosis. (The prevalence of tuberculosis in low- and middle-income countries tends to be higher—on average—than in higher-income countries. But what does that mean? In 2016, the World Health Organization estimated that there were 169,000 cases of tuberculosis in Kenya. At the time, over 50 million people were living in Kenya. While we should aim to prevent the transmission of airborne diseases like tuberculosis, statistically, it was unlikely that Sandra had the disease.) But the doctors did a comprehensive scan and suspected something else might be wrong—so they examined her uterus. At first, the physicians thought that she might have cancer. For several weeks, her thoughts were encircled with death. The sterility of the examination rooms and the mountain of paperwork in German were exhausting. Some physicians explained what was going on in English. Most did not.

Eventually, Sandra received a diagnosis—she had pleural endometriosis (a severe membrane lining in her uterus). By November 2019, five months after she arrived in Germany, the doctors recommended that she have surgery. Sandra evinces that she felt a palpable discomfort in her abdomen, a common symptom for people with endometriosis, so she assumed the surgery would impede the pain. Her medical report, which she later showed me, indicated what they did: "Transverse laparotomy, hysterectomy with salpingectomy . . . no laparoscopy, but the patient had postoperative fluid loss via the kidney." When Sandra woke up, she felt lethargic. She didn't realize until later that she no longer had a uterus. Upset, she asked the physicians why they would remove her uterus. With a laconic reply, the surgeon left little room for a thorough explanation. According to the doctors, they removed it because her "uterus was damaged." For months, everything seemed to fall apart, and she couldn't find the words for her suffering: She no longer had a uterus and did not know why. "As a doctor," Sandra bemoaned, "he was supposed to protect me or take care of me, not to damage me. If I was German, I don't think he would have removed my uterus." When the country went into lockdown in March 2020, not only did she have to challenge a medical system in a language that she did not know, but she also had to find a way to gather the resources to fight back.

It wasn't until Sandra connected to a women-led migrant group that she felt less isolated. Not only did she come across migrants who knew what it was like to leave a life behind, but she was able to communicate with other migrant women who spoke English. They helped her find a psychotherapist, connect with a lawyer, and find a sense of peace. The network gave her access to a world where people provided advice about where to buy spices and get her hair braided. She could temporarily leave the past behind and relish unspoiled laughter.

By March 2020, gathering weekly with this group came to a halt. They had virtual meetings during the lockdown and met in person in Berlin when gatherings were permitted. Now, she had to cope with the place she was living on the periphery of an East German farming town, largely sparse, surrounded by a community that felt disenchanted with the state of the world.

At the height of the lockdown, most of the shops were closed. The grocery store was the only place open. Most people in Sandra's *Lager* were careful and mostly stayed close to their bubble—immediate family. They used the kitchen one at a time and maintained a distance when they could. If they gathered, they did so outside to avoid the possibility of transmission. At first, Sandra remained in the town, but one day, after speaking to a friend, she stated, "I'm lonely." So that winter, in December 2020, Sandra visited her friend for Christmas.

Temporarily losing her connection to Berlin made her realize how much she loved the city. Not only could Sandra procure better doctors, that is, people who listened to her, but there were more Africans around. "In Berlin, now I have a lawyer to fight for my legal case." When I sat in her room, drinking a cup of warm chamomile tea, we talked about how other people coped with the pandemic at the *Lager*. When the vaccination rollout happened in the spring of 2021, most residents refused to get the shot. Some of them believed that if they were vaccinated, they could never get pregnant. Language did not help. Even when the social workers brought Arabic- and English-speaking doctors to the facility, people did not trust the process. *Why would they give them this shot but not a decision about their asylum case?* They wanted to know what exactly was going on. "I can't blame them," Sandra remarked after sipping some tea, "because we are refugees; we are going through a lot."

Trust is the thread that will determine how people perceive the vaccine. In October 2020, according to a study in *Nature Medicine*, about 70 percent of Germans indicated their willingness to take a Covid-19 vaccine if it was safe and effective. One important point to note is that there was a correlation showing that an increase in the level of mistrust of vaccines corresponded with an increase in higher rates of populism, as noted by Dr. Jonathan Kennedy, a public health researcher at the Centre for Public Health & Policy at Queen Mary University of London, and it must be noted that a shift in politics complicates vaccine hesitancy. In 2020, trust in the British government as a source of information about Covid-19 fell from 57 percent in April 2020 to 45 percent in August 2020.

One thing that keeps Sandra going is her daily conversations with her two daughters. To live in Germany as an asylum seeker far from family is to be in a state of limbo. While she waits to get official status, loved ones are dying in Kenya. "There is nothing you can do from Europe." And she's not the only one. Loss has become part of the lives of the other residents in her *Lager.* The pandemic augmented death, and their confinement in the German refugee camp curtailed bereavement.

By the summer of 2021, she began to feel more comfortable. It was easier for her to get to Berlin, and her case manager believed she might be able to transfer. Now that she regularly sought medical treatment in the capital city, life felt more suitable: "My doctors,

they are doing their best and tell me that I'm going to make it."
That's all Sandra wanted, for her doctors to do their best.

Some of Us Do Not Survive

The British novelist and playwright Deborah Levy once wrote, "To become a writer, I had to learn to interrupt, speak up, speak a little louder, and then louder, and then speak in my voice, when it's not loud at all." Ava's (this is a pseudonym) voice was delicate if speaking is the key to a literary craft. "I want to break the silence," she told me. "That is good for me. There must be a lot of sisters and brothers who want to talk about their cancer, but as a community, we don't have a group."

Ava wanted a group for other cancer survivors, African-descended people like her, who knew what it was like to work through the haptics, the medication, the nausea, the fatigue, the symptoms, but more than anything, the racism. She wanted to meet with other people who carried the onus of survival and carried empathy. When she was diagnosed with breast cancer, she was living in Frankfurt. For the first year of her chemotherapy, she did not know other Black people battling the malady. She believed that her experience as an Afro-German woman with a terminal disease was peculiar, but one thing she was proud of was that she was a survivor.

Ava had a materially safe childhood that was split between two countries. She spent the first six years of her life in Niger, surrounded by her father's family. When Ava, her mother, and her sister left the country when Ava was six, she had no idea that her family was fleeing a dictatorship. Seyni Kountché, a military dictator who led a coup d'état in 1974, had initially suspended the constitution, which banned political opposition to maintain morality and social order. Because Ava's mother was German, the family could live in Europe.

The exodus to Germany was difficult for Ava. They settled near Stuttgart in the muffled countryside for the first year. Although she never met her German grandparents, there was tension in the air. When Ava got older, she later learned that her grandparents were Nazis and that they disowned her mother for marrying her African father. When her father finally joined them, they eventually went to Göttingen, where he studied. Here, her father was part of the African Students Association and was in contact with other multinational African families. These blended families accepted her parents and sister in a way that her German grandparents had not.

"If you are silent about your pain," Zora Neale Hurston declared, "they'll kill you and say you enjoyed it." Ava refused to stay silent and rejected the stereotypes she believed mischaracterized and harmed Black women. Myths can impress an image that few people can fathom. There was only one fiction that Ava was tired of hearing: the myth of the strong Black woman. In her 1979 book, *Black Macho and the Myth of the Superwoman,* Michele Wallace argued that after the Civil Rights movement, Black womanhood was equated with inordinate stamina, which meant that they were perceived to be exempt from infirmity or pain. Although she did not cite Wallace, Ava believed that this narrative did not make space for Black women to be vulnerable—or better yet, to be cared for in the way they deserved.

Earlier in 2020, from February to April, Ava tells me that she underwent chemotherapy, a procedure that felt brusque with her body. Most people who have undergone chemotherapy are told it can damage their immune system. Typically, when a person gets a common cold, they are spared, mostly because their white blood cells have the power and strength to safeguard the body from a microbe.

However, given that chemotherapy attacks most cells—though the intention is to target malignant cells—white blood cells are reduced. An extreme infection can contribute to sepsis, a life-threatening condition that leads to the breakdown of the lungs, intestines, and other organs. Ava marked every encounter with risk, fearing that even a mild infection could compromise her infirm immune system. Covid posed a huge risk for her. For the most part, she found a rhythm in her daily routines, avoiding large crowds or even the protests she wanted to attend.

The novelist Yaa Gyasi wrote, "Some may want to call the events of June 2020 a 'racial reckoning,' but in a country in which there was a civil war and a civil rights movement 100 years apart, at some point it would be useful to ask how long a reckoning need take. When, if ever, will we have reckoned?" Gyasi implies that a reckoning may never happen, and given the slew of visible anti-Black violence carried out in the United States during the summer of 2020, many Black people—from Minneapolis to Paris—expressed their love and concern for Black life at the height of the pandemic.

Like many Black people living in Germany, Ava was devastated to see the video of George Floyd's final moments. Although racism in Germany has a different flavor, she knew that racism was not uniquely American. She was inspired and wanted to articulate her grief with others publicly. But given her compromised immune system, she stayed home for fear that she would get the novel coronavirus.

Ava was prudent at the beginning of the first Covid-19 lockdown in Germany. "For me, the stricter rules, such as mandatory mask-wearing on public transportation, are better. As a high-risk person, I feel better in public spaces. I feel like more people know how to move in public spaces. Most people wear masks." In Germany, the government has been lauded for its public health gains, albeit the surge in cases in the fall led to a countrywide quarantine on November 2, 2020, putting many parts of life on halt. Since December 27, 2020, the Robert Koch Institute—the leading public health institute in Germany—documented the German campaign, which averaged 40,000 vaccinations per day during the first week of distribution. For many of us who have taken cold walks through Berlin's dark winter, the vaccine is a reprieve from our isolation. The protesters from the fall of 2020 did not disappear. Persuading German citizens to take the vaccine was difficult, especially as there was a growing minority of people who argued that vaccines are harmful. Some skeptics saw it as part of the wider government or QAnon conspiracy, while others mistrusted the government. As journalist Katrin Bennhold reported at the time, Germany is riddled with QAnon thinkers under the banner of *Querdenken* (lateral thinking), with some of them declaring that wearing a mask is ineffective and "inhuman" (*unmenschlich*).

When Ava was initially dealing with her cancer prognosis and treatment, she was reminded of what her body had endured: The lethargy was akin to running a marathon, so her body and her daily activities often traveled at a snail's pace. During the fall of 2020, when the Covid-19 numbers began to rise in Germany, she felt lucky. Because she was immunocompromised, her employer informed her that she could continue to work from home. She initially went back and forth about whether to work from home, but when one of her coworkers tested positive for Covid-19, she was convinced to self-isolate. Ava spoke with her coworker, who informed everyone about her symptoms: Even though she was in her mid-twenties, she felt like she was constantly out of breath. *What would have happened if she had gotten sick?* She didn't mind remaining home even longer. Because of her cancer treatment, Ava had already become accustomed to retreating to her home.

After her cancer diagnosis in 1977, Audre Lorde wrote about her perpetual physical and emotional pain. All she wanted to do was survive, a feat that she likened to war. "Well, women with breast cancer are warriors, also. I have been to war and still am. So has every woman who had had one or both breasts amputated because of cancer, which is becoming our time's primary physical scourge." Ava has a different life story from Lorde, but what they share is living in the aftermath of breast cancer in Germany. After years of battling breast cancer in the United States, Lorde started spending time in Germany—most notably Berlin—during the 1980s, connecting with an Afro-German feminist community. Documenting every moment with intention, having healing circles, and forming meaningful friendships with fellow writers and activists, Lorde found sustenance and beauty. She wasn't obsessed with death. "For me, my scars are an honorable reminder that I may be a casualty in the cosmic war against radiation, animal fat, air pollution, McDonald's hamburgers and Red Dye No. 2, but the fight is still going on, and I am still a part of it."

For the first year of the Covid-19 pandemic, cancer patients, like Ava, were cautious and brooded about their risk. For people waiting for surgery or chemotherapy, their apprehension was substantiated: What if they were exposed to the virus during their hospitalization? Living in remission and daily life—taking the subway, sitting in an office, shopping for groceries—all posed a hazard for Ava. Balancing intimate connections against her health was a daily feature. After several months of avoiding closed areas, her pride swelled. She began to schedule outdoor walks with close friends, even in the middle of Germany's brutal winter. She opted to live again. When the vaccine rollout in Germany came along, she was initially reticent to get the shot. She was dubious: *What were the long-term side effects? Could we trust the efficacy?* Because of cancer and her fragile health, she was isolated as early as August 2019. Because she was anxious, she spoke with her physician. After multiple conversations, she felt some relief. Ava's doctor believed that the Covid vaccine would substantially reduce her risk of infection and severe symptoms. She could see the horizon.

"I think it is better to be protected for a certain amount than to stay at home or zig-zag through a crowd when I am in public." More than anything, she wanted physical and mental control and decided to get the Covid vaccine. After Ava got the vaccine, she felt more relaxed and more willing to go to public events. She was less worried. She was getting her monthly checkups but often worried that the cancer might return. *What if there was a surge in Covid cases and she needed cancer treatment? What if the virus mutates?* Ava missed most of her yoga classes, but she thought of returning since they have large windows, and they could ventilate the studio very easily.

For cancer survivors, there are many reasons to be wary about one's health, precisely because death's shadow looms each time the body feels like it may disintegrate. The pandemic amplified those fears. Learning to live in a coronavirus endemic world gave her a new perspective. In the fall of 2021, Ava and I returned to the myth. "If we only recount survivors' or hero stories, not everyone will relate. Not every Black person will be resilient. Who will tell the stories of those who don't survive?" For Ava, the Covid-19 lockdown had widened her perspective as she stepped gracefully into acquiescence. She could just be; she didn't have to lose herself in baseline survival.

Here Comes the Bride

In 2020, I became obsessed with reading medical reports about the novel coronavirus, Covid-19. The symptoms were alarming: loss of smell, shortness of breath, loss of taste, wheezing, a sense that your lungs would collapse, and premature death. I took all the steps to avoid the new plague and relied on the one thing I had control over—curating my social life to remain socially distant. My partner, who was an academic turned writer, and I stayed home. But how that played out was challenging and harsh to witness from my Berlin apartment. At times, we felt woeful just as much as we felt despair; we suffered from nostalgia just as much as we felt loneliness. At the same time, confinement terrified us, jostled us, and we did everything we could to escape the boredom by socializing on the internet. The pandemic had a life of its own, not because of its biological traits, but how my neighbors responded to it, and how German society contended with government-imposed lockdown. We witnessed something bigger that Covid-19 has laid bare—how history can be a dark mirror of the present.

It was devastating, to say the least, to witness the ubiquity of health inequalities among ethnic minorities, alongside the clamor of science "skeptics" who appeared to have little regard for all lives. Putting aside the more obvious xenophobia of the former US president referring to Covid-19 as the "China virus," the outright downplaying of the consequences of malady was itself a racially coded act, given the disproportionate mortality rate for non-white people in countries like the United States and the United Kingdom—where the data on racial inequities were documented, stark, and ever-present.

For many Black Americans like myself, the fear of being infected by Covid-19 coincided with the racial inequities I observed from my computer screen. From midwestern cities like Detroit and Milwaukee to semi-rural communities in Alabama and Louisiana, Black Americans were dying in disproportionate numbers from the novel coronavirus. One study found that in Chicago, where 30 percent of the population is African American, they accounted for 70 percent of all coronavirus deaths. These statistics were chilling to read, but they reminded me why I left the United States in 2017 and why building a life abroad meant finding some form of normality that could displace the confinement I had grown accustomed to.

The pandemic accelerated my full membership into adulthood. In August 2020, I married my partner on a whim. We eloped to Copenhagen—the Las Vegas of Europe—for a ceremony attended by two. Living in Berlin we were accustomed to the sclerotic bureaucracy of Germany, the litany of procedures that often left even the most basic tasks requiring a trip to the postal office, a German-engraved stamp, and a civil servant who takes great pleasure in upholding their power by verifying your identity. Denmark had none of these perilous, paper-heavy trails.

So, we prepared for our wedding in another country, in a city three hundred miles from the home we created. To get our license, we scanned our American and British passports, filled out an online application, paid a fee, and waited for their response. A few days later, we received an encrypted email. We sat on his couch and my partner unceremoniously opened the document—luckily, it wasn't spam. We had been approved to get married.

My parents couldn't join us, as US citizens were banned from traveling to Europe at the time. His parents were reluctant to travel to continental Europe because of the logistical hurdles. Visitors traveling to Denmark from Europe had to show proof of a reservation for six days of accommodation in the country—a measure the Danish government had presumably introduced to reduce tourism without killing it off altogether. There was also a generalized cloud of anxiety around holding a public gathering, even in the summer and outdoors; a vague sense of guilt hung over the whole concept. The summer was full of stories of super-spreader weddings and parties which went viral, where brides, grooms, partners, or relatives were left with life-threatening diseases or even death. In the end, none of our relatives or friends attended our covert wedding.

We were two people who found ourselves in love as witnessed by a Danish government official. When we took a seven-hour train from Berlin to Copenhagen, I had just completed my second round of IVF, so my hormones were untamed, bountiful, unmoored, and dotted on my face. I was buoyed by the novel trip but ruffled by the uncertainty. As we crossed the border, we were asked for our passports and proof of our negative Covid-19 test—something that became a regular feature of Western European life that summer.

The six-day rule meant that we decided to begin our trip with four nights in a small cabin in the woods near Vejby, thirty-one miles north of Copenhagen. Resting in the wooded small valley less than two hundred yards from the Scandinavian Sea, we grilled seasonal vegetables, hiked along the rocky beach, and enjoyed the somnolent village. My husband-to-be even swam in the icy waters of the Kattegat, where the North and Baltic Seas meet. It was nothing, he boasted, for someone brought up in the North of England. For that weekend, in this seaside town, we had forgotten about the moment we lived in, the pandemic that had plunged our lives into a constant state of social seclusion. Free to frolic in nature, we were briefly unimpeded, and vestiges of our pre-pandemic euphoria resurfaced.

A few days later, we were in Copenhagen, preparing for our wedding. Sensible people who planned their lives would have had wedding rings. I am an impulsive and free-spirited person, and my partner can fall easily into disorganization and chaos when fashion is concerned. As such, we didn't acquire bands until the day before. Given this, we searched for rings that matched our desperation and budget—we went to a secondhand shop in Copenhagen and found an undersized ring for my partner. None of the rings were of my liking so I decided to use an aqua-colored ring I brought with me on this trip. We received a steady set of calls and texts from our loved ones congratulating us on our big day.

Would it be bad luck to wish someone a good wedding before it actually happened? Would I be a runaway bride for the civil service without an audience?

These questions floated through my head, fuzzy doubts that had more to do with my ongoing anxiety than anything real.

If there was anything we lacked, it was following form and ritual. As we prepared ourselves for the ceremony in our accommodations, we not only spent the night together, we prepared our attire on the top floor of a minuscule loft apartment in the city. My partner was fiercely fashioning himself in his best image—he moisturized his face with a skincare beauty regimen I taught him—primarily because of the fear that his aging skin would be more visible well before mine. I mostly prepared myself from the mid-century wooden chair, adding a light foundation, eyeliner, mascara, and lipstick to my face. As I put on my leggings I nearly fell over and I knocked over some of the art books in the apartment. His dark blue suit was sculpted for his body and my silver dress, designed by a Berlin-based Cameroonian seamstress, made me feel like I was wrapped in an intergalactic puff. Our mood was deliberately gentle, and we maintained a quiescent satiety, drifting into a place of contentment.

Fifteen minutes before we left for Copenhagen City Hall, I began to bleed. My expression fell to the floor. Two weeks before, a fertility center had inserted two fertilized embryos—my egg and my partner's sperm—into my uterus. This was our second reproductive assistance attempt; I thought that this time we would succeed. When my partner embraced me, I broke down and a stream of tears fell on my cheeks, smudging some of my eyeliner and mascara along the way. After a brief lament, I had to change gears. The wedding needed to be portentous of a new future with my husband-to-be. After nearly half a year of cohabitation during a pandemic, we had defied all of the odds, become each other's support during the most difficult moments of isolation.

Like the five other couples at the Copenhagen City Hall we saw when we arrived on a balmy summer day, our skins glistening against the celestial luminescence. We yearned to declare our love for a ten-minute secular ritual, in a town hall that was alluring with its elaborate frescoes and gilded statues. I didn't have to walk down an aisle or recite a vow. Instead, the officiate—a portly man with a cherub face—asked us to confirm our names. When we verbally validated our identities, he proceeded with the ceremony. He read us a secular Danish love poem and asked us to sign the marriage certificate. My partner rushed to the task eagerly declaring that he would accept me as his wife. Being the indecisive person that I am, I hesitated. Long enough to reflect, but not so extensive that it was awkward. I looked up at his face, saw his smile, and scribbled my esoteric signature on the form. Shortly after, the official congratulated us on our marriage. By this point, my heart was racing so much that all I could do was kiss my partner to slow down the pace.

Were we actually married? Why did I suddenly get a rush of joy?

We were taken aback by the fact that although our ceremony had been curtailed by the presence of the virus, we were both tremendously moved by our newfound union. Precautions demanded a ritual that could not be shared with our friends or family. A "safer" Covid-19 marital union meant a wedding of two, with two government-appointed officials, masked up and intimate.

Throughout the days, walking hand in hand, wearing our smiles and our ersatz rings—strangers on the streets of Copenhagen congratulated us. Our joy infected them, and we were in turn astonished that they cared—thinking that this would never happen in bacchanal Berlin, where many people prided themselves on non-marital polycules. Our wedding was an exception to the life we had been living for the last several months—here, on the streets of Denmark, we were drowning in public displays of affection. We were lucky. For one afternoon in the summer of 2020, we forgot the weight of the pandemic. We didn't think of the pre-existing conditions that left some people vulnerable to infection or the medical failures of the United States. We were free, and our nuptials felt like a contradiction to the quandaries in the world, as we were used to, and the world, which had been transformed.

Even though we were relatively safe during the initial year of the Covid-19 pandemic—we had jobs that allowed us to work from home, we had secure housing, and we were relatively young and healthy—the Covid-19 lockdown emotionally famished and exhausted us. We had become so accustomed to our social bubbles, social distancing, and social isolation that we had forgotten the unbridled joy that brought us together.

That spring, in March 2020, I had moved out of my apartment in a former ladder factory—living with a collection of German and European leftists, a disorganized but prodigious apartment that had functioned as a great party space—with my partner into a standard Berlin apartment. I grieved my former life, often flooded with loads of people. Since the spring of 2020, I've felt disconnected from the feminists who organized art shows, late night dancing at the queer techno clubs that became zones of revelry, and some of the best falafel sandwiches in town. My then-neighborhood of Neukölln was and continues to be a surrogate home for many of the misfits that I know. So, leaving this district left a gap in my world. My partner grieved his single life and adjusted to a boisterous, disheveled roommate—me. I lost some friendships. He missed his family. We were mourning our past lives, even if we celebrated each other. And our wedding day, even if yet another fertility failure stained it, was a visible reminder of our love, our joy, unfolded for the public, visible to ourselves and these strangers in Copenhagen.

I soaked in the lines from Paulo Coelho's *The Alchemist*: "When we love, we always strive to become better than we are. We strive to become better than we are; everything around us becomes better too." Expressing our adoration freely and publicly was in contradiction to what we had been living with. Social distancing had, in effect, become social isolation, and I managed to forget what it was like to sit with this type of public freedom. The constant cycle of loss, grief, elation, and disappointment wasn't unique to me. Countless friends I spoke with in New York City, London, Paris, and Istanbul were shaken and broken. We were on edge because the world as we knew it was about as fickle as a toddler deciding what toy to play with.

Retreating to Nature

In the fall of 2021, I found myself on the edge of a verdant cliff, where the forest was hugging the sea. Here, the clouds gliding steadily across the sky appeared so low that I could reach out and touch them. After two hours of walking along the French Mediterranean coast, my afternoon was colored by a panoply of evergreen shrubs resting on limestones and a vivid display of cacti overlaid with grisly spikes. With each step, I grew accustomed to the curviness of juniper trunks jettisoning from the tawny carpet of dust. Even if I was staring absentmindedly at the sea or annoyed by the chorus of mosquito hums, I drew my attention to the surroundings and the refined elements of the space. Behind every shrub, there were airborne insects; in front of every stone, there were fallen leaves. Although I was alone, I was reminded that the terrain was alive when I heard the sonic palette of seagulls. In the summer of 2021, the second year of the Covid-19 pandemic, as I walked along the cliff, I was breathless from the hot and muggy air and arrested by the commanding briny smell of the sea.

I felt like an inmate that absconded from jail. Here, I found a better part of my mind and I was no longer confined. Nestled between Marseille and Cassis, the Calanques are a series of steep, almost sheer-sided valleys formed by fluvial erosion and scattered with caves, a geology whose most notable features are the caverns and underground streams. As I continued this walk, I moved through the aqua-colored swimming holes and green shrubs scattered on a rugged ledge so close to the sea. At the same time, I was bewitched by this expansive, mysterious space where this beauty whistled with a tune, accessible and pouring out of the landscape. This is where I found some reprieve: full-bodied, organic, healing. For all the sensations that trickled inside me, I was allured—not knowing why—by the scintillating clouds. More than anything, the forest gave me the space to face my history—a Haitian-descended person walking through rural France, aware that the country's wealth and landscape were augmented through the exploitation of my ancestors. Knowing the history while surrounded by a wild landscape left me captivated with unapologetic wonder while calling into question what it means to be a steward of the Earth.

I was temporarily residing in Cassis, a small town once known for its fish, wine, and limestone quarries, now more famous for tourism, attending a residency that gave me the space and time to read and write next to the dramatic congregation of waves. So close to Marseille, a cosmopolitan city home to people of Maghrebian, West African, and Caribbean descent, I expected to see more people of color during my hikes. I was especially struck by how Marseille has and continues to feature as a point of inspiration for Black writers, who sketched out the quays of the Vieux Port or gave sordid accounts of the nightlife. Claude McKay's posthumously published novel *Romance in Marseilles* captures the city's pulse: "All the jazz hounds who raised hell in the mighty cities of earth were summoned here by the Almighty to welcome him. All the saints were strutting their stuff, and the angels fluttered their wings for him, the center of attraction." McKay's Marseille was a meeting spot for the subaltern and the salacious. Still, his work is nearly absent from the Calanques and other nearby natural areas. Even when Afro-Caribbean people like Claude McKay found themselves abroad in France, very close to nature, they were often excluded from the landscape around Marseille, in some cases confined in its inner neighborhoods.

Walking through the Calanques during the late summer of 2021, I was caught by the sonic rituals of nature and reveled in the subtle differences in flora. Still, I also felt the solemn weight of being the only Black American stumbling, breathing, and walking through these European landscapes. Even in bliss, I absorbed the incongruity of entering this atmosphere, feeling misaligned and misunderstood, and trying to channel these periodic bursts of catharsis by opting out of the digital pandemic sacraments of online conference calls and entering the stillness of the countryside.

The countryside reminded me that the years of the pandemic constrained me emotionally and caused me to store up so much grief.

The African American historian Keeanga-Yamahtta Taylor noted in her essay "The Black Plague" that "this macabre roll call reflects the fact that African-Americans are more likely to have pre-existing health conditions that make the coronavirus particularly deadly." These inequalities—whether hastened by Covid-19 or police brutality—became a perennial stain in my life. Before I went to Cassis, I didn't know how much my body was holding, how my emotions were bottled to a fault. I was carrying the weight of collective loss and survivor's guilt, consistently working and opening us up to the inexorable pull of writing full-time. This felt like a chilling predicament as if I were setting myself up for failure. Some days, I thought about what it meant to write from the position of the working class, the precarity of life, or even the souls of Black folk.

The coronavirus years have led to a new search for ways to be expansive and to escape, and so on. In the summer of 2022, my partner and I went on a secular pilgrimage, walking part of the Chemin Saint Jacques, a route that leads ultimately to Santiago de Compostela in northwestern Spain. For eight days, we hiked and camped in this southern French landscape: settling into a familiar rhythm, rising before dusk, drinking repugnant instant coffee, and drowning in our perspiration from the prolonged heat wave. When lucky, our early afternoons were colored by a regalia of pine trees that could provide some reprieve from the ascending heat wave. But as my partner and I hiked across this landscape, we encountered a drought, an arid landscape of bereft French towns with a vivid display of dust overlaid with grisly foliage. The terrain was alive, though, rustling with the sounds of leaves, the breathless greetings of our fellow walkers, and the sonic palette of cows expressing their discontent with our invasion. I felt both constrained by the oppressive heat and gloriously free under an open sky.

We became pilgrims, invading the delicate crevices of nature, foliage-strewn, seemingly frozen in time. Moving through the crest of the plateaus and the troughs of the valleys, we celebrated, with glee, when we were shielded from the sun's rays by towering trees. At the same time, the expansive, mysterious space bewitched me, especially in moments during an arduous climb. We were not religious, but our actions ran parallel to the miserly practices of the pious. By exercising material prudence, we were depriving ourselves of bodily indulgences, mainly because we were hiking by carrying everything we needed on our backs. Each step reminded us of what we needed to get through the world and what we were missing. When we camped alongside various riverbanks and forests, we fed each other, not just through food, but by finding the language to think about the mountains, forests, and trees as an extension of our humanity. More than that, I wanted to understand how pushing my body could lead to transcendence—would I reach a higher state of consciousness if I felt the pain throbbing through my knees?

As a hiker, I tried to reflect on how we related to each other. As we passed through the French countryside in the searing heat, I wondered what we owe the Earth. In her influential work *Primate Visions,* the feminist scholar Donna Haraway asks, "What forms does love of nature take in a particular historical context? For whom and at what cost?" This question reveals that our connection—or lack thereof—to nature is not an arbitrary event but tied to the past and present circumstances that constitute our reality. Loving the environment, as Haraway puts it, is contingent on time and place, dependent both on how humans perceive their relationship to nature and, in a way, how the heart responds to humans. But my hike compelled me to think more deeply about my character, my resilience, and the pandemic. I reflected on the family members I lost through death and the friends I lost from aloofness. Being unrestrained meant that I was stirred by the landscape and reckoning with what it means to move through the world easily. But most importantly, it gave me the sense to reacquaint myself with what I had been searching for, the freedom to write.

"Once you find yourself in another civilization," Baldwin noted, "you're forced to examine your own." Wherever we live in the world, Black Americans could not escape the collective grief that overwhelmed our community after George Floyd, a Black man, was murdered by a white police officer in Minneapolis. But the grief wasn't just tied to his death under police custody; it was connected to the never-ending injustice in Covid-19 deaths, and the economic oppressions that left many people of African descent underpaid or in debt. Apart from a student loan of $140,000, the trauma of Black life keeps me away from the United States. My family, too, has been affected by the American carceral state, which preys on Black people. In 2014, my brother—who suffers from schizophrenia—was incarcerated for more than eight months for trespassing on someone's lawn, and in 2019, he was shot and survived a gun wound. During the first year of the pandemic, one of my cousins was imprisoned in Florida, awaiting his release in solitary confinement to avoid contracting Covid-19. In February of 2024, that cousin died of a brain aneurysm at the age of forty.

Like James Baldwin, who fled America's suffocating racism and spent some pivotal years living in Europe during the Civil Rights movement, I once felt contrite about living abroad during a pivotal moment when there appeared to be a reckoning. From Berlin, I was afforded the space to write, breathe, and be debt-free.

The privilege to write and regard the land as a space of leisure suggests that implanting ourselves into the Earth—even when most of us did not historically have the authority to write about ourselves—is manumission. What helped me survive in the moments when I felt confined under the lockdown was my composition. In my mind, the loss was a constant.

The Covid-19 pandemic showed that we all need freedom, but depending on whom you asked, it meant different things. For QAnon supporters, they equated liberty with the right to frequent a certain business or to cough unimpeded in the face of a fellow commuter on a city bus. For people concerned about their safety, confinement was mixed, but we also knew that before Covid-19, some of us were already constrained. One lesson I learned was that stress was overwhelming, living emotionally on a knife's edge. We needed space and time to ourselves.

Chapter 7

LOCKED UP

Of course, writing novels about the future doesn't give me any special ability to foretell the future. But it does encourage me to use our past and present behaviors as guides to the kind of world we seem to be creating. The past, for example, is filled with repeating cycles of strength and weakness, wisdom and stupidity, empire and ashes. To study history is to study humanity. And to try to foretell the future without studying history is like trying to learn to read without bothering to learn the alphabet.

—Octavia E. Butler

FOR HAITIANS, WHO INSIST ON being heard in their own words, uploading a video through social media can generate an electric, viral response. One reminder of the power of these digital narratives came from a video uploaded to Facebook in June 2022. The recording is shaky and most likely made with a smartphone. As the camera moves through a poorly lit and overcrowded room, several men begin to emerge in the background. A voice, which appears to be that of the cameraman, is speaking Haitian Creole outside the frame.

"Me li oui. Ou we bagay sa wi. Me glo a, gad cule glo a."
(There it is; you see that right. Here's the water; look at its color.)

He continues to describe the scene: the meager food, the dirty floor, and the broken windows. And here, another speaker interjects to state: "They don't give us anything." They have no mattresses, the men in the video claim; they are supplied only with cardboard boxes to sleep on. As the video continues to pan around the space, we see men lying on the floor, surrounded by fraying plastic bags. There are nineteen men in this facility, the narrator indicates. Pointing to the concrete ceiling, where shards of dirty-white paint peel away, the narrator says, "If there is an earthquake, this is where we will die." There is no toilet; the men have a hole in the ground to relieve themselves.

Near the end of the clip, someone enters the room and tells the videographer to calm down. The man retorts, "I can't calm down because I am suffering." At that point, the camera halts.

Russian literature isn't something I think about very much, day to day. Still, in the fall of 2022, during a period when discourse and public health measures concerning Covid-19 began to wane, infectious outbreaks were still prevalent in prisons. When I received a flurry of videos from the Haitian prison, I was reminded of Leo Tolstoy's novel *Resurrection:* "All these institutions [prisons] seemed purposely invented for the production of depravity and vice, condensed to such a degree that no other conditions could produce it, and for the spreading of this condensed depravity and vice broadcast among the whole population." Haiti in the twenty-first century is far removed from nineteenth-century Russia, but one similarity was very evident; prisons are more concerned with depriving people of their humanity than "solving" social problems.

The video from the National Penitentiary was posted in June 2022 by Haitian journalist Theriel Thelus. It was not just the content that disturbed me, but the conditions that led to the men's incarceration. Although the Haitian Ministry of Justice has a fiduciary responsibility to provide prisoners with details of their arraignment and trial, these men, like 81 percent of incarcerated people in Haiti, had yet to be charged with a crime. According to the United Nations Human Rights Council, Haiti routinely sees "the illegal and arbitrary use of pre-trial detention." In some cases prisoners remained incarcerated because they could not afford to pay for their release, echoing the fate of imprisoned people in the United States who wait for months, or sometimes even years, in jail in appalling conditions because they cannot pay for bail.

The prisoners' prolonged detention was made all the more shocking by the overcrowding. The official capacity of the Haitian prison system is only suitable to contain over 2,000 individuals; there were approximately 11,000 people incarcerated. Not only did the conditions leave prisoners little room for privacy or to rest, but overcrowding, poor ventilation, and unsanitary water created ideal conditions for them to get sick. As bad as the circumstances were in the summer of 2022, by October they were even worse—the prison became the locus of a resurgent cholera epidemic in Port-au-Prince, which left 16 prisoners dead in the first week of October. In the National Penitentiary, where the outbreak was at its most severe, the Haitian journalist Widlore Mérancourt remarked, a facility designed for 800 inmates housed 4,000 people.

"What does it mean to defend the dead?" Christina Sharpe wrote. "To tend to the Black dead and dying: To tend to the Black Person, to Black people, always living in the push toward our death? It means work."

On January 12, 2010, Haiti was hit by a huge earthquake, which killed over 250,000 people and left over one million Haitian people homeless. That summer, I decided on a whim to move to Port-au-Prince to help with relief efforts. As a recent graduate of a public health master's degree program, and a person of Haitian descent, I thought my expertise and linguistic skills could be helpful to the survivors.

The situation was grave when I arrived there. Amid endemic poverty and political instability, and with almost no remaining infrastructure, the people of Haiti were struggling to rebuild their lives. Things got significantly worse a few months later when a new killer emerged on the scene: cholera.

The cholera epidemic broke out near a military base housing the United Nations security force. Between October 2010 and March 2011 the epidemic infected over 800,000 Haitians and killed over 10,000 of the country's population. With the island's primary water source, the Artibonite River, being contaminated, public health efforts were further crippled. This was the first large-scale cholera outbreak in the twenty-first century, yet it did not shock many in the Global North. After all, what happened in Haiti wholly matched their expectations of epidemics—deadly outbreaks of ancient diseases affecting foreign people in exotic, faraway, and poor lands. Well-meaning Europeans and North Americans were saddened by the news coming from the Caribbean Islands; some even donated money and supplies to try to help. But most did not see the outbreak as a wake-up call to review the readiness of their own countries or the global community for a similar epidemic or pandemic.

Meanwhile, I was living in a working-class neighborhood in Port-au-Prince, distressed and anxious, worried that I would get cholera. The dilemma of witnessing people who looked like me— Black individuals—subjected to premature death left me broken. I tried to quell my anger about how working poor, darker skinned Haitians were subjected to the burden of ill health, but I could not. So I left Haiti, and since then, I have not returned.

When I began receiving reports of the cholera outbreak in Haiti in 2022, I was reminded why I had left a decade earlier. "To die from cholera is a nightmare both to experience and to witness," a quartet of physicians and public health practitioners wrote in *The Nation*, "as Haitian communities described during the first cholera outbreak." The perennial presence of death can leave a scar, but my fear this time had more to do with how I felt when I left, swallowed by remorse and wondering what it would mean to suppress my personal anxieties and choose to stay for the collective good. I set out to connect with Haitian people living in the country in atonement for my abrupt departure twelve years before.

During the summer of 2022, I made contact with Saguens Bernabe, a Haitian-born man in his early thirties who a few months earlier had been deported from the United States, where he spent part of his childhood and all of his adulthood. In his videos, messages, and calls, he communicated something more than the dread and quiet lassitude I associated with the unsanitary conditions of the prison. Like other deportees who spoke to the media during this time, such as Patrick Julney, who was wrongfully detained by the Haitian authorities, Bernabe was inflamed not only over his own experiences; he had a burning desire for justice for the men he had been imprisoned with, who had suffered from the October 2022 cholera outbreak. Bernabe had been set free in August 2022, but he continued to advocate for those still inside.

Corresponding with Bernabe reminded me of so many past conversations with Haitians. They have a way of telling their stories, a gentleness that subtly reveals their past, dispensing clues as to where they fit into our broad diaspora. During one conversation, I asked him if he still spoke Haitian Creole. He let out a peal of light laughter and remarked that he knew most things but would be lost on some words. His language didn't progress so far after he moved to the United States.

Bernabe migrated to Florida in 1997 at the age of nine. Able to secure a visa for travel, he, his mother, and his brother joined his father in Delray Beach, a small city fifty miles north of Miami. Like many Haitians who emigrated during the late nineties, his family sought to escape the surge in political insecurity and violence. His story is similar to that of thousands of childhood arrivals, people who spent most or part of their childhood in the United States. His early life experience was, in many ways, similar to mine, but it differed in at least one crucial respect. Born in Miami, I am a US citizen. Bernabe traveled on a "family-sponsored" visa and secured permanent residency.

Deportations of Haitians increased markedly after Donald Trump's government withdrew their temporary protected status. The Dominican Republic persistently deports people, too, sometimes on mere—racialized—suspicion that people with darker complexions are automatically Haitian, and not Dominican. Even US permanent residents such as Bernabe were not secure; conviction for a serious crime renders one liable to deportation. He fell short thanks to a minor infraction: possessing a firearm and marijuana in his car.

While living in Akron, Ohio, in 2018, he and a friend purchased a firearm at a gun show. In Ohio, firearms may be openly carried in cars only with a permit, which Bernabe admits he did not have. On top of that, his friend had some marijuana with him while Bernabe was driving. Both charges fell on Bernabe because it was his car. He was arraigned with improper handling of a firearm in a motor vehicle and drug possession. While the state of Ohio has somewhat decriminalized marijuana, this applies only for medical purposes. Bernabe was indicted with a misdemeanor and sentenced to three years in prison, of which he served two years. After his release, he sought advice from his lawyer, requesting a citizenship application, but that flagged immigration authorities, which considered his misdemeanor a deportable offense.

Despite his appeal, Bernabe was detained and by early 2022, he was expelled from the United States.

Bernabe did not deny fault with the law as it stood, nor did he express anger at his arrest, but I could not help but think that the forces that shape American policing had a lot to do with his deportation. In a 2020 report, the Stanford Open Policing Project found what most African Americans know: Black drivers are more likely to be stopped by the police when compared to white drivers. Once stopped, they are more likely to have their vehicles searched. With twenty million motorists stopped by police annually, there are thousands of opportunities for a Black man to be caught committing a minor, non-violent offense of the kind that for another person would go undetected. If a person isn't a US citizen, like Bernabe, the consequences can be life-changing, and utterly devastating.

Bernabe was one of at least thirty Haitian men who were partially raised in the United States and deported in the first half of 2022. Some were deported after receiving a prison sentence, others for being undocumented. None of them expected their deportation to a country most of them had not been to since they were children. "You are not allowed to get any of your luggage," Bernabe told me in a calm voice. "The clothing you have on your back is pretty much the clothing you will do for the week."

Bernabe arrived in Port-au-Prince on April 19, 2022, landing at the Toussaint Louverture International Airport, where he was turned over to the Haitian authorities. Although he and the other deportees expected to be picked up by family members, they waited over a fortnight in a holding cell unsure if and when their family members would arrive. On May 6, 2022, he was told that he would be released, expecting to meet a maternal cousin whom he had not seen in over a decade. Instead, he was taken to the Haitian National Penitentiary.

The journey lasted just ten minutes, but Bernabe describes it like a scene in an action film, as if he and the other deportees were high-profile criminals who might escape at any moment. The police officers drove briskly, weaving onto the wrong side of the road, not stopping, while the prisoners remained shackled, holding on to what they could for dear life, hoping they would make it.

Ohio was difficult, boring, and sometimes dangerous, but landing in a Haitian prison was far worse. Guards mostly left the men to their own devices. On a typical day, Bernabe describes, "You have to wait until your family members bring you some food for today or some water."

"Were there people there who didn't eat at all," I asked, "or were there other ways people would find food?"

"When that happens, we share food with them."

Fictional versions of prison, such as the TV show *Oz*, often depict men in their most hardened state. Indifferent to others, scorning any form of care. We are led to believe that they have to fend for themselves, that they have to combat their incarceration alone or align themselves with ethnocentric gangs. The accounts revealed that the men displayed emotions far beyond simple hypermasculinity and overt violence. The men demanded to be seen, to be heard, for their humanity to be acknowledged; they were trying to show each other, both inside and outside, how they were mistreated.

Being detained is a form of psychological and physical torment, Homer Venters writes in his book *Life and Death in Rikers Island*. The condition makes one susceptible to disease. This situation extends far beyond Rikers. Bernabe became ill multiple times while incarcerated: He suffered from fever and diarrhea, and his ankles swelled up. There was a small medical staff in the prison, but they rarely prescribed anything. He had been vaccinated for Covid-19, but he was wary of the fact that there were few precautions, such as mask wearing or social distancing. In truth, the latter would hardly have been possible. In the worst cases of sickness, the prison authorities would take a prisoner's blood pressure or check their vitals. Once the first cases of cholera started emerging at the prison, things began to change. At the time of the outbreak, Bernabe had been released, but he remained in touch with some of the men inside through Facebook and WhatsApp. He was saddened and enraged when his friend Roody Fogg died in October of what others inside described as cholera symptoms. Bernabe lamented Fogg had been forced to return to Haiti from New Jersey, only to die.

When Bernabe was released four months after his arrival in Haiti, on August 30, 2022, it wasn't through the benevolence of the Haitian state or even a trial; it seemed to be mere luck. A tenacious family member kept going to the courthouse asking the same questions: What law did Bernabe break? What infractions was he being held for? According to Bernabe, his cousin irritated the authorities until they finally gave up and released him. Yet many more should be free. As the *Haitian Times* has remarked, relatives in the United States and Haiti have demanded the release of men jailed at the National Penitentiary, with many protesting outside the General Consulate of Haiti in New York City. Bernabe surmised that he would still be in prison if not for his cousin's tenacity—a distressing alternative, as Haitian prisons deal with a period of intense disorder themselves. As Tanvi Misra reported for *The Nation*, being confined in Haiti at present means being "caught in [a] crossfire of gang violence inside and outside the prison walls, and it entails endangering distant family and acquaintances by asking them to wade through gunfire and protests to drop off food or medication." Knowing how intractable the scourge of violence engulfing Haiti has been, and the lax administrative measures in place to care for those subjected to its carceral state, I asked Bernabe how he kept up his stamina given the odds. He responded, "I don't feel right. I got out and the rest of them are still in there. And sure enough, after I started speaking up, some of the guys started getting released."

In November 1970, shortly after the FBI captured Angela Davis for alleged involvement in a shooting at a Marin County courtroom, the novelist James Baldwin wrote a letter to her at the Marin County Women's Detention Center in California, highlighting the intolerable conditions that shaped Black American life inside and outside prison walls. "If we know," he wrote, "then we must fight for your life as though it were our own—which it is—and render impassable with our bodies the corridor to the gas chamber. They will come for us that night if they take you in the morning." Somewhere in Baldwin's message is what Bernabe sought to achieve: Freedom for oneself means trying to liberate all those in chains.

His efforts were not in vain, and with each person's release, there were more people to fight from the outside. By the end of 2022, most of the thirty deportees were released from the Haitian National Penitentiary. When people at the margins, such as the incarcerated, strive toward their liberation as individuals, justice isn't always resolute. However, when they work collectively, they might find some retribution. Most people have been released, but they remain separated from their families and former lives. It is unclear whether they will see their US homes again. Even out of jail, or out of the video frame, they are a reminder that there is no reason to be calm when you're suffering.

If public health measures are to be an effective weapon against the current pandemic, and those to come in the future, it will mean upending the system that has caused these drastic inequalities. Public health measures applied universally to an uneven and unequal world will not work. That world must be radically rearranged. Public health policy requires up-to-date and peer-reviewed research that considers how epidemics impact all populations, exercising empathy for those who are disenfranchised, and understanding how we all fit into the complex social systems that shape our life paths.

The 2022 cholera outbreak in Haiti took its shape at a level much bigger than the individual and was born in an institution—the National Penitentiary—that was indelibly marked by the society that built and maintained it. But perhaps there is a lesson to be taken into account: the robust durability of the Haitian people's spirit for resilience and capacity to rebuild after turmoil. As the novelist Edwidge Danticat has written in the preface of René Philoctète's novel *Massacre River,* Haitians are "as fluid as the water themselves," adapting wherever they are, whether they want to or not.

Postscript

REFLECTING IN "AIDS AND ITS Metaphors," Susan Sontag intoned: "Epidemics of particularly dreaded illnesses always provoke an outcry against leniency or tolerance—now identified as laxity, weakness, disorder, corruption: unhealthiness. Demands are made to subject people to 'tests,' to isolate the ill and those suspected of being ill or transmitting illness, and to erect barriers against the real or imaginary contamination of foreigners." Sontag describes how various states enforced quarantines and periods of detention for people who were suspected of being vectors of HIV/AIDS. She points out how contemporary public health measures, implicitly and explicitly, confined "at risk" groups—sex workers, the urban poor, foreigners—in language and practices that are a portent of war.

One can confine marginalized people precisely because their death may not elicit the same empathy as the death of a member of the elite, and because if, still living, their life is diminished one can count on the same invisibility. The passing of an incarcerated Haitian man, for instance, starts to feel "natural," while the end of a young white girl in the United States is a tragedy. This calculus bears no direct relationship to how diseases spread, at the level of microbes and symptoms. Instead, it speaks to the places, times, and manner in which public health officials mobilize their efforts. Whom do we save, and on what terms do we do so?

Epidemics have become an avatar for the expression of our pre-existing fears; in a period of crisis, all our psychoses emerge. "Not only did the [Covid-19] pandemic cripple the existing political system," Masha Gessen observed in *Surviving Autocracy*, "but it also demanded that [President] Trump governs in precisely the manner to which he aspired: unilaterally, decisively, with few checks on his power—and with the eyes of the nation riveted to him." That leader, Trump, ruled as the spear of an authoritarian, corporatized culture that managed to reap benefits of a global lockdown and mass confinement. It further opened a window to ride on people's anxiety as they responded to the coronavirus in the ways they wanted to see themselves—resilient, competent, scared, confused. A mix of emotional responses, rooted in a mentality of threat and scarcity, that all too often result in further depreciation of the already undermined.

In January 2023, the US government announced that people from Venezuela, Cuba, Nicaragua, and Haiti crossing the US-Mexico border would be returned to Mexico and denied asylum. The message was clear: Migrants from these countries were not welcome. Title 42, a public health provision, granted the US government the temporary ability to expel immigrants suspected of having a communicable disease. Under Trump the Covid-19 pandemic became a pretext for asylum seekers to be denied entry into the United States. Upon Title 42's suspension under President Joseph Biden in May 2023, a surge of migrants seeking refuge appeared at the southern border of the United States.

What made Title 42 (and other US migration rules) so pernicious is that it sets health—or a distorted picture of it—as a standard marker of fitness for those wishing to exercise a right to move to and within the United States. These immigration policies warp the meaning of humanity, as the philosopher Sylvia Wynter observed: "Being human, in this context, signals not a noun but a verb." Though fear of infectious disease is often reasonable and of genuine concern, quarantines, lockdowns, and exclusions alone cannot eliminate the spread of illness if most of the population lacks access to free healthcare, stable housing, and wholesome food. In contrast, such emergencies may act as a distraction from the root causes of our society's ills; the same afflictions that exacerbate the impacts of those emergencies in the first place.

The United States exceeded over 1.2 million Covid-related deaths as of April 2024. One cannot separate these deaths from the glaring reality that 33.2 million people in the United States remain uninsured. The movement for Medicare for All is a loose coalition of millions of Americans who have demanded a single-payer universal healthcare system. Yet, time and again, their efforts have fallen short, not for lack of trying, but because of the relentless inertia of the American political system. This cannot be taken into account without a look at the rightward shift in American politics and the deadly consequences of late-stage capitalism.

What does it mean that we have become immune to the deaths of the poor, the dispossessed, and the disenfranchised? This is the moral question that I am still working through, as I try to make sense of the histories told here, of microbes and people.

As I wove between the past and the near present, my own encounters with confinement braided into these case studies. These are snapshots of history, of moments in which people reckoned with death. I wanted to know what made people sick, what motivated them to find some reprieve—whether medicinal or psychological—and to understand better how people exercise care. The impulse, so often present, to detain, confine, and incarcerate, revealed something about how we saw each other and how we saw ourselves. Confinement also came with the exploitative labor; it could be used to conduct experiments, too. We have become accustomed to a world where forced captivity is justified for the disenfranchised, with a judicial system deciding that someone needs to be removed from society. I want us instead to ask how we can challenge the conditions that condemn certain people to premature death.

In the absence of reconciliation for the aggrieved, I find assurance in community-oriented health collectives—including the Black Panther Party social programs, not associated in everybody's knowledge with a practice of care—that stand against the consumer logic of healthcare that has become ubiquitous in the United States. I am with those who advocate for what some have called "health communism," proposing a radical re-thinking of health that moves away from the language of scarcity, and does away with the vocabulary of "undeserving" and "deserving."

Medical knowledge is abundant, yet we are told these resources should be rationed rather than shared with collective solutions that dissolve the racial and class disparities of health. When people are detained at the border or in prison, shuttered away into slums, excluded from society, and deprived of the opportunities to flourish, we are reminded that these forms of captivity lead directly to poor health. Since 2020, we have all had to grapple publicly with new fears and novel experiences of contagion, and navigate bereavement in our quieter, private states. This internal reflection remains an incomplete project. So far, I have reached a simple but potent conclusion: Gratuitous death is not merely a product of confinement, rather it happens when a section of society is neglected and bereft of aid. Some of us survived, but over the years, I have learned that survival is not enough.

Acknowledgments

THIS BOOK IS AN OFFERING to the many people who assisted me along the way and the multiple debts I have incurred. In this list of gratitude, I must begin with the fact that this book would never have seen the light of day without the invaluable team at One Signal for their editorial support and my agent, Ian Bonaparte, whose guidance was instrumental in bringing this book to fruition. No work of history is possible without the support of the librarians, archivists, and institutions that make original research possible. For that, I thank the Max Planck Institute for the History of Science, Ludwig Maximilian Universität in Munich, the Camargo Foundation, Baldwin for the Arts, the Robert B. Silvers Foundation, and the Andy Warhol Foundation for the Visual Arts for providing an institutional home during the writing and editorial stages.

This project was written during the first few years of the COVID-19 pandemic, but it began years before, during my quest to understand how the most vulnerable are shaped by illness. I am indebted to the archivists who directed me and the subjects interviewed, especially those who revealed intimate details about confinement. A group of people read and commented on all or significant parts of the book as it was being written and revised. These include Adrian Chen, Ben Mauk, Carleen Coulter, Christienna Fryar, Harry Stopes, Jared Malsin, and Jessica Loudis. I also want to thank my editors for other publications—Atossa Araxia Abrahamian, Chloe Stead, Clare Longrigg, David Marcus, Eric Sullivan, Jess Bergman, Mariya Petkova, Terence Trouillot, and Vanessa Peterson for providing me the opportunity to develop my literary voice.

Beyond this support, this work would not be possible without my comrades and collaborators who have been generous with their time: Adam Benkato, Akua Gyamerah, Alice Spawls, Ayasha Guerin, Ben Miller, Caoimhe McAlister, Charmaine Li, Christine Kikare, Daniel Trilling, Diana Kwon, Duane Jethro, Emilia Roig, Erica Cardwell, Hanno Hauenstein, Heba Gowayed, Jennifer Wilson, Kemi Fatoba, Lamia Moghnieh, Lisa Onaga, Luiza Prado, Melody Howse, Morgan Jerkins, Natasha Kelly, Nathan Ma, Nnenna Onuoha, Ruth Michelson, Sarah Jaffe, Sarah Shin, Thomas Turnball, Yasmina Price, and Zoé Samudzi.

All my achievements are the direct result of the unconditional love and encouragement of my family: my parents, my cousins, my sister, and my brother. This book could not have been completed without my partner, Harry Stopes, for his care and unwavering support. I continue to be inspired by him and our son for their strength and bravery.

Endnotes

PROLOGUE

xi *"Everyone who is born ... citizens of that other place":* Susan Sontag, *Illness as Metaphor and AIDS and Its Metaphors* (New York: Penguin Books, 1991), 3.

xi *"The history of illness ... interest of a few":* Anne Boyer, *The Undying: A Meditation on Modern Illness* (New York: Penguin Books, 2019), 30.

xiv *fourth largest cause of death:* Centers for Disease Control and Prevention, *Morbidity and Mortality Weekly Report* 48, no. 15 (October 1999): 905–932, https://www.cdc.gov/mmwr/PDF/wk/mm4840.pdf.

xiv *Eighty percent of them:* Jonathan H. Mermin et al., "Typhoid Fever in the United States, 1985–1994," *Archives of Internal Medicine* 158, no. 6 (March 1998): 633–638, https://doi.org/10.1001/archinte.158 .6.633.

xv *"Most of the unpleasure ... as a 'danger'":* Sigmund Freud [Translated by James Strachey], *Beyond the Pleasure Principle* (New York: W. W. Norton & Company, 1961), 5.

xvi *"While the state ... margins of political life":* Michel-Rolph Trouillot, *Haiti, State Against Nation: The Origins and Legacy of Duvalierism* (New York: Monthly Review Press, 1990), 16.

xvi *drowning their passengers:* Gregory Jaynes, "33 Haitians Drown as Boat Capsizes off Florida," *New York Times*, October 27, 1981, https: //www.nytimes.com/1981/10/27/us/33-haitians-drown-as-boat-capsizes

-off-florida.html#:~:text=Thirty-three%20Haitians%2drowned%20
this,detention%20facility%20for%20illegal%20aliens.

xvii   *the basis of their nationality:* Jean William Pape et al., "The Epidemiology of AIDS in Haiti Refutes the Claims of Gilbert *et al.*," *Proceedings of the National Academy of Sciences* 105, no. 10 (March 11, 2008): E13, https://doi.org/10.1073/pnas.0711141105.

xix   *"We tell and feel ... black life":* Katherine McKittrick, *Dear Science and Other Stories* (Durham, NC: Duke University Press, 2021), 9.

xx   *Spanish is by far the most common:* Brittany Shammas, "Miami is Only U.S. City Where Most Language Learners Are Studying English," *Miami New Times,* October 13, 2017, https://www.miaminewtimes.com/news/miami-is-only-us-city-where-most-studied-language-is-english-9743341.

xxi   *impacted African American health:* Dorothy E. Roberts, *Killing the Black Body: Race, Reproduction, and the Meaning of Liberty* (New York: Pantheon Books, 1997).

xxi   *"trying to ameliorate ... probably futile, approach":* Harriet Washington, *Medical Apartheid: The Dark History of Medical Experimentation on Black Americans from Colonial Times to the Present* (New York: Anchor Books, 2006).

xxi   *"The* something *... is racism":* Linda Villarosa, *Under the Skin: The Hidden Toll of Racism on American Lives and on the Health of Our Nation* (New York: Knopf, 2022).

xxi   *write in their book* Inflamed*:* Rupa Marya and Raj Patel, *Inflamed: Deep Medicine and the Anatomy of Injustice* (New York: Farrar, Straus and Giroux, 2021), 49.

xxii   *a 2022 study in Italy:* Damian Carrington, "Microplastics Found in Human Breast Milk for the First Time," *The Guardian,* October 7, 2022, https://www.theguardian.com/environment/2022/oct/07/microplastics-human-breast-milk-first-time#:~:text=feed%20a%20baby.-,The%20breast%20milk%20samples%20were%20taken%20from%2034%20healthy%20mothers,on%20living%20humans%20remains%20unknown.

xxii *life expectancy has fallen:* Robert H. Shmerling, "Why Life Expectancy in the US is Falling," *Harvard Medical School,* October 20, 2022, https://www.health.harvard.edu/blog/why-life-expectancy-in-the-us-is-falling-202210202835.

xxiv *his seminal work* Freedom*:* Orlando Patterson, *Freedom: Volume 1: Freedom in the Making of Western Culture* (New York: Basic Books, 1991).

xxv *the novelist Arundhati Roy wrote:* Arundhati Roy, "The pandemic is a portal," *Financial Times,* April 3, 2020, https://www.ft.com/content/10d8f5e8-74eb-11ea-95fe-fcd274e920ca.

xxvii *"Illness is the night-side life":* Sontag, *Illness as Metaphor,* 3.

CHAPTER 1: CONTAGION ON THE PLANTATION

1 *"The hold of slavery . . . and the dispossessed":* Saidiya Hartman, "The Hold of Slavery," *New York Review of Books,* October 24, 2022, https://www.nybooks.com/online/2022/10/24/the-hold-of-slavery-hartman/.

2 *he referred to as the "prognathous race":* Samuel A. Cartwright, *Ethnology of the Negro or Prognathous Race* (New Orleans, LA: s.n., 1857).

2 *Cartwright alleged that Black people:* Cartwright, *Ethnology of the Negro.*

3 *In his 1832 book:* Samuel A. Cartwright, *Some Account of the Asiatic Cholera, Cholera Asphyxia, or Pulseless Plague; with a Sketch of Its Pathology and Treatment from the Best Authors, and Some Original Remarks; Also, Advice Relative to Its Prevention on Plantations, and Its Mitigation, Premonitory Symptoms and Treatment, Should It Occur* (Natchez, MS: The Natchez, 1833).

3 *slave cabins "should be aired and kept clean":* Cartwright, *Some Account of the Asiatic Cholera.*

4 *"whatever may be . . . gives it subjects":* Cartwright, *Some Account of the Asiatic Cholera.*

5 *"on plantations . . . to mitigate its violence":* Cartwright, *Some Account of the Asiatic Cholera.*

5 *One planter's medical companion:* James Ewell, *The Planter's and Mariner's Medical Companion: Treating, According to the Most Successful*

Practice, I. The Diseases Common to Warm Climates and on Ship Board. II. Common Cases in Surgery, as Fractures, Dislocations, &c. &c. III. The Complaints Peculiar to Women and Children. To Which Are Subjoined, a Dispensatory, Shewing How to Prepare and Administer Family Medicines, and a Glossary, Giving an Explanation of Technical Terms (Philadelphia, PA: John Bioren, 1807).

6 *"... cure or injure a negro":* Samuel A. Cartwright, "Report on the Disease and Physical Realities of the Negro Race," *New Orleans Medical and Surgical Journal,* May 1851, 691–715.

6 *he prescribed the woman:* Stephen N. Harris, "Case of Ovarian Pregnancy," *Southern Journal of Medicine and Pharmacy* 1, no. 1 (January 1846): 371–77.

6 *Cartwright's recommendation for treating cholera:* Samuel A. Cartwright, *Some Account of the Asiatic Cholera.*

7 *Achille Mbembe evinces that:* Achille Mbembe, *Necropolitics* (Durham, NC: Duke University Press, 2019), 16–17.

7 *Clifton Ellis and Rebecca Ginsburg assert that:* Clifton Ellis and Rebecca Ginsburg, *Cabin, Quarter, Plantation: Architecture and Landscapes of North American Slavery* (New Haven, CT: Yale University Press, 2010).

8 *As historian Rana Hogarth noted:* Rana A. Hogarth, *Medicalizing Blackness: Making Racial Difference in the Atlantic World, 1780-1840* (Durham, NC: University of North Carolina Press, 2017), 1–2.

9 *Black captives resided in a cabin:* James Clifton, *Life and Labor on Argyle Island* (Savannah, GA: The Beehive Press, 1978).

10 *Louis Manigault, the owner, boasted that:* Louis Manigault, Plantation Manual, April 24, 1860.

10 *Louis Manigault wrote to his father:* Manigault, Plantation Manual, December 26, 1854.

11 *"Considering the immense losses ... in the midst of harvest":* Manigault, Plantation Manual, May 1, 1856.

12 *the author Susan Sontag tells us:* Sontag, *Illness as Metaphor,* 37.

13 *With Louis Manigault noting:* Manigault, Plantation Manual, May 1854.

13 *he expressed sanguine comments:* Louis Manigault, "Respecting the Cholera amongst the People in Nov[ember] & Dec[ember] 1852," January 1, 1853, Louis Manigault Papers, Special Collections Library, Duke University.

13 *described by Manigault as:* Manigault, "Respecting the Cholera amongst the People."

13 *given that he believed:* Manigault, Plantation Manual, June 18, 1862.

14 *According to his records:* Louis Manigault, "Prescription Book," Louis Manigault Papers, Special Collections Library, Duke University, 1852.

15 *in June 1850 alone:* US Census Bureau, "United States Tables, The Seventh Census June 1, 1850," accessed April 5, 2024, https://www2 .census.gov/library/publications/decennial/1850/1850b/1850b -03.pdf.

15 *listed in the census by their enslaver:* US Census Bureau, "History: 1850," accessed April 5, 2024, https://www.census.gov/history/www /through_the_decades/index_of_questions/1850_1.html.

16 *he concluded that a microbial agent:* "Microscopic Observations and Pathological Deductions on Asiatic Cholera," *The British and Foreign Medico-Chirurgical Review* 16, no. 31 (1855): 144–45, https://www .ncbi.nlm.nih.gov/pmc/articles/PMC5182670/.

17 *Professor Saidiya Hartman asserts:* Hartman, "The Hold of Slavery."

18 *took matters into his own hands:* Virginia Jayne Lacy and David Edwin Harrell, "Plantation Home Remedies: Medicinal Recipes from the Diaries of John Pope," *Tennessee Historical Quarterly* 22 no. 3 (September 1963), 259–265.

19 *decided to concoct a "remedy":* Lacy, "Plantation Home Remedies."

19 *Mr. Pope claimed that:* Lacy, "Plantation Home Remedies."

19 *Ewell argued that cholera:* James Ewell, *The Planter's and Mariner's Medical Companion: Treating, According to the Most Successful Practice, I. The Diseases Common to Warm Climates and on Ship Board. II. Common Cases in Surgery, as Fractures, Dislocations, &c. &c. III. The Complaints Peculiar to Women and Children. To Which Are Subjoined, a Dispensatory, Shewing How to Prepare and Administer Family Medicines, and a*

Glossary, Giving an Explanation of Technical Terms. Philadelphia: John Bioren (Philadelphia, PA: Printed by John Bioren, 1807), 141.

19 *Dr. Ewell recommended consuming:* Ewell, *The Planter's and Mariner's Medical Companion.*

20 *"That the constitution ... possibly afford him":* Ewell, *The Planter's and Mariner's Medical Companion.*

20 *The cost of a physician:* Virginia Jayne Lacy and David Edwin Harrell, "Plantation Home Remedies: Medicinal Recipes from the Diaries of John Pope," *Tennessee Historical Quarterly* 22 no. 3 (1963): 259–65, https://www.jstor.org/stable/42621637.

20 *In his memoir:* Frederick Douglass, *My Bondage and My Freedom: An African American Heritage Book* (New York: Penguin Books, 2003 [1855]), 39.

21 *Speaking of Uncle [Doctor] Isaac Cooper, Douglass remarks:* Douglass, *My Bondage and My Freedom,* 58.

22 *Linda Brent:* Harriet Jacobs, *Incidents in the Life of a Slave Girl* (New York: Penguin Books, 2000 [1861]).

22 *Harriet described Black maternal mortality:* Jacobs, *Incidents in the Life of a Slave Girl,* 16.

23 *she blamed his death:* Jacobs, *Incidents in the Life of a Slave Girl,* 232–233.

23 *"He peopled . . . my master":* Jacobs, *Incidents in the Life of a Slave Girl,* 30.

23 *Jacobs wrote:* Jacobs, *Incidents in the Life of a Slave Girl,* 43.

23 *she chided:* Jacobs, *Incidents in the Life of a Slave Girl,* 43.

24 *She "often prayed ...":* Jacobs, *Incidents in the Life of a Slave Girl,* 63.

24 *with "chills and fever":* Jacobs, *Incidents in the Life of a Slave Girl,* 68.

25 *read with and against the archive:* Saidiya V. Hartman, *Scenes of Subjection: Terror, Slavery, and Self-Making in Nineteenth-Century America* (Oxford, UK: Oxford University Press, 1977), 80.

26 *As Angela Davis notes:* Angela Y. Davis, *Women, Race, & Class* (New York: Vintage Books, 1981).

26 *successful births within the slave community:* Deirdre Cooper Owens,

Medical Bondage: Race, Gender, and the Origins of American Gynecology (Athens, GA: University of Georgia Press, 2017).

26 *worth as much as 25 percent more:* W. E. B. Du Bois, *Black Reconstruction in America: Toward a History of the Part Which Black Folk Played in the Attempt to Reconstruct Democracy in America, 1860-1880* (New Brunswick, NJ: Transaction Publishers, 2013).

27 *"At least death frees the slave from his chains":* Jacobs, *Incidents in the Life of a Slave Girl*, 258.

27 *"...slave woman ought not to be judged...":* Jacobs, *Incidents in the Life of a Slave Girl*, 62.

27 *births resulted in maternal mortality:* Shannon K. Withycombe, "Women and Reproduction in the United States during the 19th Century" (Oxford, UK: Oxford University Press, January 25, 2019), https://doi.org/10.1093/acrefore/9780199329175.013.426.

27 *"Healthy pregnancy was hardly possible...":* Dorothy E. Roberts, *Killing the Black Body: Race, Reproduction, and the Meaning of Liberty* (New York: Pantheon Books, 1997), 47.

28 *her desire to escape:* Jacobs, *Incidents in the Life of a Slave Girl*, 129.

28 *"You are my slave...":* Jacobs, *Incidents in the Life of a Slave Girl*, 67.

29 *McKittrick writes about this conundrum:* Katherine McKittrick, *Demonic Grounds: Black Women and the Cartographies of Struggle* (Minneapolis, MN: University of Minnesota Press, 2006), 40–41.

30 *Hippocrates wrote in his treatise:* Hippocrates, *Of the Epidemics: Book 1* (Berlin: De Gruyter Akademie Forschung, 2014).

30 *her confidence thickened:* Jacobs, *Incidents in the Life of a Slave Girl*, 222.

31 *"...by means of an education":* Du Bois, *Black Reconstruction in America*.

32 *One congressional report stated:* "Report of the Commissioner of the Bureau of Refugees, Freedmen and Abandoned Lands" [BRFAL], in U.S. Congress, House Executive Documents, 39th Cong., 1st sess., no. 11, 18 (serial 1255); George R. Bentley, *A History of the Freedmen's Bureau* (Philadelphia, PA: Univ. of Pennsylvania, 1955), 76.

32 *Rev. A. S. Fiske:* John Eaton, *Grant, Lincoln, and the Freedmen* (New York: Longman's Green, 1907).

35 *recalling the large volumes of patients:* Rebecca Crumpler, *A Book of Medical Discourse* (Boston, MA: Keating, 1883), https://collections .nlm.nih.gov/catalog/nlm:nlmuid-67521160R-bk.

36 *"If you stick a knife . . . admit the knife is there":* Malcolm X, "If You Stick a Knife in my Back," YouTube video, https://www.youtube.com /watch?v=XiSiHRNQlQo.

37 *Canadian scholar Rinaldo Walcott tells us:* Rinaldo Walcott, *The Long Emancipation: Moving toward Black Freedom* (Durham, NC: Duke University Press, 2021), 1.

37 *". . . I was not comfortless . . .":* Jacobs, *Incidents in the Life of a Slave Girl*, 128.

CHAPTER 2: THE AFRICAN LABORATORY

41 *Figure 1.* Robert Koch and Stabsarzt Kleine with a dead crocodile, East Africa. Process print after a photograph, 1906/1907, Wellcome Collection, https://wellcomecollection.org/works/yzg3fjhg.

42 *Robert Koch asked the German microbiologist:* Friedrich Karl Kleine, *Ein deutscher Tropenarzt* (Hannover: Schmorl & von Seefeld, 1949), 44.

42 *"bacteriology is a very young science . . .":* Robert Koch, "On Bacteriology: And its Results, A Lecture," Delivered at the first General Meeting of the Tenth International Medical Congress, Berlin, August 4, 1890.

43 *problem for Europeans colonizing Africa:* Robert Koch, "Koch to the Minister der Geistlichen, Unterrichts- und Medizinalangelegenheiten, Berlin, 30 Juli 1904," in *Gesammelte Werke*, ed. Robert Koch, Georg Theodor August Gaffky, J. Schwalbe (Leipzig: Thieme, 1912), 926.

44 *"He brought . . . from Bukoma":* Robert Koch, "Report on the Sleeping Sickness Expedition during the Stay in Muansa," July 31, 1906, Robert Koch Papers, Images 1593–1605, Staatsbibliothek zu Berlin, Berlin, Germany.

45 *create a controlled environment:* Robert Koch, "Report on the Activities of the Sleeping Sickness Expedition Sese Entebbe and Planskizze,"

November 25, 1906, Robert Koch Papers, Image 1618–1630, Staatsbibliothek zu Berlin, Berlin, Germany.

45 *caused weight loss and blindness:* Dietmar Steverding, "The Development of Drugs for Treatment of Sleeping Sickness: A Historical Review," *Parasites & Vectors* 3, no. 1 (March 10, 2010): 15, https://doi.org/10.1186/1756-3305-3-15.

45 *"A concentration camp exists . . .":* Andrea Pitzer, *One Long Night: A Global History of Concentration Camps* (New York: Little, Brown and Company, 2018), 5.

46 *several hundred cases per year:* World Health Organization, "Trypanosomiasis, Human African (Sleeping Sickness)," May 2, 2023, https://www.who.int/news-room/fact-sheets/detail/trypanosomiasis-human-african-(sleeping-sickness).

47 *definition of who could be German:* George Steinmetz, *The Devil's Handwriting: Precoloniality and the German Colonial State in Qingdao, Samoa, and Southwest Africa* (Chicago, IL: University of Chicago Press, 2007).

47 *As Helen Tilley notes:* Helen Tilley, *Africa is a Living Laboratory: Empire Development and the Problem of Scientific Knowledge, 1870–1950* (Chicago, IL: University of Chicago Press, 2016), 12.

49 *Speaking of his motivations for science:* Robert Koch, "Journal of Outdoor Life," 5 (1908), 165 –169.

49 *saw his vocation as an act:* Robert Koch, "On Bacteriology: And its Results, A Lecture," delivered at the first General Meeting of the Tenth International Medical Congress, Berlin, August 4, 1890.

50 *the archives showed:* Robert Koch, "Ätiologie der Milzbrandkrankel, begründet auf die lung, Entwicklungsgeschicte des Bacillus Anthracis," *Beträiäge zur Biologie Planzen*, 2 (1876): 27–310.

52 *citing an 1896 epidemic:* Robert Koch, Letter from Robert Koch to the minister of instruction and medical affairs, Berlin, July 30, 1904, Robert Koch Papers, Bl. 146–150, Staatsbibliothek zu Berlin, Berlin, Germany.

52 *asserting in July 1904:* Robert Koch, Letter from Robert Koch to the minister of instruction and medical affairs, Berlin, July 30, 1904, Rob-

ert Koch Papers, Bl. 146–150, Staatsbibliothek zu Berlin, Berlin, Germany.

52 *approved Koch's request:* The president of the [German] imperial health office, September 17, 1904, Robert Koch Papers, Transcript III. B. 4521, Reference number 7347/04, Staatsbibliothek zu Berlin, Berlin, Germany.

53 *50,000 Marks for other costs:* The president of the [German] imperial health office, September 17, 1904, Robert Koch Papers, Transcript III. B. 4521, Reference number 7347/04, Staatsbibliothek zu Berlin, Berlin, Germany.

53 *"After the* Reich Health Council . . .":* Wolfgang U. Eckart, "The Colony as Laboratory: German Sleeping Sickness Campaigns in German East Africa and in Togo, 1900–1914," *History and Philosophy of the Life Sciences* 24 no. 1 (2002): 69–89. Also refer to Wolfgang U. Eckart, *Medizin Und Kolonialimperialismus: Deutschland 1884–1945* (Paderborn: Schöningh, 1997).

55 *"They gave . . . modern genocidal ideologies":* Achille Mbembe, *Necropolitics* (Durham, NC: Duke University Press, 2019), 24.

56 *speaking of patients:* Robert Koch, "Final Report on the Activities to Study Sleeping Sickness," *Deutsche Medizinische Wochenscrift* no. 46 (April 25, 1907), 543.

57 *"A village has been . . .":* Robert Koch, Report on the activities of the sleeping expedition Sese Entebbe and Planskizze, November 25, 1906, Robert Koch Papers, Staatsbibliothek zu Berlin, Berlin, Germany.

58 *placed into these camps:* Robert Koch, "Mitteilung i.iber den Verlauf und die Ergebnisse der vom Reiche zur Erforschung der Schlafkrankheit nach Ostafrika entsandten Expedition, personal report, Reichsgesundheitsrates," November 18, 1907 (Chair: Prasident Dr. Bumm), in *Gesammelte Werke* vol. II, part 2, ed. J Schwalbe (Leipzig: Thieme, 1912), 930–940, 936.

58 *No. 168:* Robert Koch, Report on the activities of the sleeping sickness expedition up to November 25, 1906, Bericht Sese Entebbe, Robert Koch Papers, Staatsbibliothek zu Berlin, Berlin, Germany.

58 *No. 236:* Koch, "Report on Sese Entebbe and Planskizze."

59 *No. 527:* Koch, "Report on Sese Entebbe and Planskizze."

59 *"Koch's main task ...":* Manuela Bauche, "Robert Koch, die Schlafkrankheit und Menschenexperimente im kolonialen Ostafrika," Freiburg Postkoloniale, https://www.freiburg-postkolonial.de/Seiten/robertkoch.htm.

60 *"The improvement in the condition . . .":* Koch, "Final Report," 534–546.

60 *his report makes no mention:* Koch, "Final Report," 534–546.

60 *the results for humans:* Koch 1912d, 537–538.

61 *Koch demanded quarantine:* Koch, "Final Report," 534–546.

62 *Walter Rodney provides receipts:* Walter Rodney, *How Europe Underdeveloped Africa* (Washington, D.C.: Howard University Press, 1981), 189.

63 *In his journal, he described:* Friedrich Karl Kleine 1949, 44; Genschorek 1981, 192. [In the original German, they use the term *Mikroskopierzelt.*]

64 *his 1907 essay:* Koch, "Final Report," 544.

65 *Ehrlich injected:* Robert Koch R. 1912e, "Uber den bisherigen Verlauf der deutschen Expedition zur Erforschung der Schlafkrankheit in Ostafrika," in *Gesammelte Werke,* ed. Robert Koch, Georg Theodor August Gaffky, J. Schwalbe (Leipzig: Thieme, 1912).

66 *doctors also forged connections:* Deborah J. Neill, *Networks in Tropical Medicine: Internationalism, Colonialism, and the Rise of a Medical Specialty, 1890–1930* (Stanford, CA: Stanford University Press, 2012).

66 *A 1910 essay:* Claus Schilling, "Welche Bedeutung haben die neuen Forstchritte der Tropenhygiene für unsere Kolonien," in Verhandlungen des deutschen Kolonialkongresses (Berlin: Verlag Kolonialkriegerdank, 1910), 162–85.

66 *prominent role in documenting:* Claus Schilling, "Die Schulen für Tropenmedizin in England," *Klinisches Jahrbuch* (1907): 495–501.

66 *alleged these medical trials:* Hans Ziemann, "Über die Schlafkrankheit in Großkamerun," *Beihefte zum Archiv für Schiffs- und Tropenhygiene* 16 (1912): 112–140.

67 *conducted experiments:* Hans Ziemann, "Is Sleeping Sickness of the

Negroes an Intoxication or an Infection?," *Journal of Tropical Medicine* 5 (October 15, 1902): 309–14.

67 *the high malaria rate:* Hans Ziemann, "Wie erobert man Afrika für die weisse und farbige Rasse? Vortrag gehalten auf den Internationalen Kongress für Hygiene und Demographie zu Berlin, 1907," *Archiv für Schiffs- und Tropenhygiene* 11 (1907).

67 *part of a colonial system:* Ludwig Külz, "Guinée française und Kamerun," *Amtsblatt für das Schutzgebiet Kamerun* 13–16 (1909): 115–18, 133–44, 144–48, 163–68.

70 *modern biology emerged:* Angela Saini, *Superior: The Return of Race Science* (New York: Penguin Random House, 2020).

CHAPTER 3: WHO'S AFRAID OF THE FLU?

75 *Figure 2:* S. Burt Wolbach and Channing Frothingham, "The Influenza Epidemic at Camp Devins in 1918: A Study of the Pathology of the Fatal Cases," *Arch Intern Med (Chic)* 32, no. 4 (1923): 571–600.

76 *wrote to his friend:* Dr. Roy from Camp Devens, A Letter from Camp Devens, September 29, 1918, https://www.pbs.org/wgbh/american experience/features/influenza-letter/.

76 *a mild economic recession:* François R. Velde, "What Happened to the U.S. Economy during the 1918 Influenza Pandemic? A View Through High-Frequency Data," *The Journal of Economic History* 82, no. 1 (2022): 284–326.

76 *A December 1918 article:* Unknown Author, "Bulletin of the Woman's Medical College of Pennsylvania," 69 (December 1918).

77 *by exercising efficiency in care:* Ellen S. More, "A Certain Restless Ambition: Women Physicians and World War I," *American Quarterly* 1, no. 4 (December 1989): 636–60.

77 *officials noted to the public:* "The Spanish Influenza," *New York Times*, October 7, 1918, https://timesmachine.nytimes.com/timesmachine /1918/10/07/97031378.html?pageNumber=12.

78 *the "Cough of Perinthus":* Georgios Pappas, Ismene J. Kiriaze, and Mat-

thew E. Falagas, "Insights into Infectious Disease in the Era of Hippocrates," *International Journal of Infectious Diseases* 12, no. 4 (2008): 347–50.

77 *observations of the epidemic:* "Notice of the Influenza 1803," *Annals of Medicine* 2 (1802): 479–486, https://pubmed.ncbi.nlm.nih.gov/30 299718/.

79 *lethal strain swept across Europe:* Tom Quinn, *Flu: A Social History of Influenza* (London: New Holland Publishers, 2008).

79 *recounts in his article:* P. Berche, "The Enigma of the 1889 Russian Flu Pandemic: A Coronavirus?," *Presse Med.* 51, no. 3 (September 2022): 104111.

79 *"In some cases . . . medicine for patients":* Yorkshire Evening Post, May 8, 1891.

79 *killed nearly 100,000 British people:* Yorkshire Evening Post, May 8, 1891.

80 *ran an article:* The New York Herald, January 31, 1890.

81 *Manchester Evening News:* 13 November 1918, https://www.british newspaperarchive.co.uk.

82 *officials were reluctant:* Mark Honigsbaum, *Living with Enza: The Forgotten Story of Britain and the Great Flu Pandemic of 1918* (London: Macmillan, 2009), 122–23.

83 *a salacious article:* "Triple Murder and Suicide: An Attack of Influenza," *Times* (London), November 6, 1918.

83 *doctors occasionally cited psychosis:* Sir William Osler, with Thomas McCrae, *The Principles and Practices of Medicine*, 8th ed. (New York and London: D. Appleton, 1915), 118.

83 *deteriorated the nervous system:* George H. Savage, "The Psychosis of Influenza," *Practitioner* 52 (January–June 1919): 36–4.

83 *a link between influenza and mental health:* Karl A. Menninger, "Psychoses Associated with Influenza," *Journal of the American Medical Association* 72, no. 4 (January 25, 1919): 236–40.

83 *"stranger to themselves":* Sigmund Freud, *The Basic Writings of Sigmund Freud*, Translated and edited by Dr. A. A. Brill (New York: Random

House, 1995); Rachel Aviv, *Strangers to Ourselves: Unsettled Minds and the Stories That Make Us* (New York: Farrar, Straus and Giroux, 2022).

84   *London was still haunted:* "The Great Death," *Times* (London), February 2, 1921, 11.

84   *backlog of corpses to bury:* British National Archives, Ministry of Health, 20/22, November 6, 1918.

86   *she reflected on her pain:* Virginia Woolf, *The Diary of Virginia Woolf: Volume 1: 1915-1919*, ed. Anne Oliver Bell (Orlando, FL: Houghton Mifflin Harcourt, 1977), 165.

86   *In July 1918, she recalled:* Woolf, *The Diary of Virginia Woolf: Volume 1*, 163.

86   *Woolf nonchalantly remarks:* Woolf, *The Diary of Virginia Woolf: Volume 1*, 165.

87   *described the illness with abnegation:* Woolf, *The Diary of Virginia Woolf: Volume 1*, 317.

87   *attending to a sick person:* Virginia Woolf et al., *On Being Ill* (Ashfield, MA: Paris Press, 2012), 57.

88   *In a cavalier remark:* Woolf, *The Diary of Virginia Woolf: Volume 1*, 24.

88   *Writing to her friend:* Virginia Woolf, Virginia Woolf's letter to Violet Dickinson, New York Public Library, Berg Collection, September 1904.

89   *did not believe her work:* Woolf, *The Diary of Virginia Woolf: Volume 1*, 9.

89   *"I thought I was probably dying…":* *The Letters of Virginia Woolf, Volume 2*, Letters from Woolf to Vanessa Bell: December 24, 1919, 407; January 2, 1920, 411.

89   *writing was shaped by illness:* Janine Utell, "View from the Sickroom: Virginia Woolf, Dorothy Wordsworth, and Writing Women's Lives of Illness," *Life Writing* 13, no. 1 (2016): 27–45.

89   *through economical prose:* Woolf, *The Diary of Virginia Woolf: Volume 2*, 132–133.

90   *"Melancholy diminishes … too much of myself":* Virginia Woolf, *Virginia Woolf: A Writer's Diary* (San Diego, CA: Harcourt, 1954), 28.

90 *contracted influenza again:* Woolf, *The Diary of Virginia Woolf: Volume 2,* 156.

91 *"not a word has been recorded ...":* Woolf, *The Diary of Virginia Woolf: Volume 2,* 176.

91 *losing her teeth:* Woolf, *The Diary of Virginia Woolf: Volume 2,* 176.

91 *her seminal essay:* Virginia Woolf, *On Being Ill.*

92 *"illness has not taken ...":* Woolf, *On Being Ill.*

92 *"... the divine relief of sympathy":* Woolf, *On Being Ill.*

94 *"There was an emptiness ...":* Virginia Woolf, *Mrs. Dalloway* (New York: Penguin Books, 2021).

95 *about trauma and recovery:* Karen DeMeester, "Trauma and Recovery in Virginia Woolf's Mrs. Dalloway," Modern Fiction Studies 44, no. 3 (Fall 1998): 649–73.

95 *"Too many have died ...":* Katherine Anne Porter, *Pale Horse, Pale Rider* (New York: Harcourt, 1979), 269.

96 *"Fiction is not fact ...":* Thomas Wolfe, *Look Homeward, Angel* (New York: Penguin Classics, 2016 [1929]).

96 *Emre articulated the significance:* Merve Emre and Virginia Woolf, *The Annotated Mrs. Dalloway* (New York: W. W. Norton, 2021).

97 *"... the barbarian within":* Susan Sontag, *Illness as Metaphor,* 63.

97 *inveighed against the pandemic plot:* Alexander Alter, "The Problem with the Pandemic Plot," *New York Times,* February 20, 2022, https://www.nytimes.com/2022/02/20/books/pandemic-fiction.html.

98 *transcribed in her journal:* Audre Lorde, Cancer Journals (January 26, 1979).

99 *succinctly put it:* Carolyn Lazard, "How to be a Person in the Age of Autoimmunity," *Cluster Magazine,* January 2013.

99 *"From my hospital bed ...":* Lazard, "How to be a Person in the Age of Autoimmunity."

100 *the significance of writing from bed:* Leah Piepzna-Samarasinha, *Care Work: Dreaming Disability Justice* (Vancouver, Canada: Arsenal Pulp Press, 2018).

101 *sequenced the influenza genome:* Christopher F. Basler et al., "Sequence

of the 1918 Pandemic Influenza Virus Nonstructural Gene (NS) Segment and Characterization of Recombinant Viruses Bearing the 1918 NS Genes," *Proceedings of the National Academy of Sciences* 98, no. 5 (2001): 2,746–51.

101 *kills up to 650,000 every year:* World Health Organization, "Influenza," October 3, 2023, https://www.who.int/news-room/fact-sheets/detail/influenza-%28seasonal%29.

102 *Her suicide note:* Virginia Woolf, "Suicide Note," Smith College Archives, date accessed April 5, 2024, https://www.smith.edu/woolf/suicidewithtranscript.php.

CHAPTER 4: BREAKING THE WALLS OF SILENCE

105 *expressed that Boudin was:* David Gilbert, *Love and Struggle: My Life in SDS, the Weather Underground and Beyond* (Oakland, CA: PM Press, 2012).

107 *shot and killed two police officers:* James Feron, "Kathy Boudin Given 20 Years to Life in Prison," *New York Times,* May 4, 1984, https://www.nytimes.com/1984/05/04/nyregion/kathy-boudin-given-20-years-to-life-in-prison.html.

107 *"Boudin is not a sympathetic figure . . .":* Sydney H. Schanberg, "New York: The Boudin Trial," *New York Times,* February 28, 1984, https://www.nytimes.com/1984/02/28/opinion/new-york-the-boudin-trial.html?searchResultPosition=6.

107 *"Many of us are looking . . . to make these real":* Kathy Boudin, "Teaching and Practice," *Harvard Educational Review* 63, no. 2 (1993): 207–233.

108 *African American women in New York State:* Federal Centers for Disease Control, *HIV/AIDS Surveillance Report* (October 1994); New York City Department of Health, *AIDS Surveillance Update* (October 1994).

108 *had more AIDS cases than any other state:* New York State Commission on Correction, 1988.

109 *being denied treatment:* Sarah Schulman, *Let the Record Show: A Politi-*

cal History of ACT UP New York, 1987–1993 (New York: Farrar, Straus and Giroux, 2021), 242.

109 *"Women were dying . . .":* Mireya Navarro, "Conversations: Katrina Haslip; An AIDS Activist Who Helped Women Get Help Earlier," *New York Times,* November 15, 1992.

110 *her motivation for co-founding ACE:* The Women of the ACE Program, *Breaking the Walls of Silence: AIDS and Women in a New York State Maximum-Security Prison* (New York, New York: The Overlook Press, 1998), 156.

111 *85 percent . . . had children:* Maryann Bird, "The Women in Prison: No Escape Stereotyping," *New York Times,* June 23, 1979, https://www .nytimes.com/1979/06/23/archives/the-women-in-prison-no-escape -from-stereotyping-the-bottom-of-the.html.

111 *sentenced to fifteen years in prison:* The Women of the ACE Program, *Breaking the Walls of Silence,* 158.

111 *"ACE changed my life . . .":* The Women of the ACE Program, *Breaking the Walls of Silence,* 159.

113 *"Instead of seeing AIDS . . .":* Kathy Boudin, "Teaching and Practice," *Harvard Educational Review* 63, no. 2 (1993): 207–33.

113 *Their publication,* Alert to *AIDS*: Boudin, "Teaching and Practice," p. 223.

114 *". . . where margins can become centers":* Lorraine O'Grady, "Olympia's Maid: Reclaiming Black Female Subjectivity," in *The Feminism and Visual Culture Reader,* ed. Amelia Jones (London: Routledge, 2003), 174–87.

114 *"there is no glory in punishing":* Michel Foucault, *Discipline & Punish: The Birth of the Prison* (New York: Random House, 1995), 10.

116 *one and a half million people incarcerated:* "Number of Persons und the Supervision of Adult Correctional Systems in the United States, 2010– 2020," Bureau of Justice Statistics, March 10, 2022, date accessed April 5, 2024, https://bjs.ojp.gov/featured/report-number-persons -under-supervision-adult-correctional-systems-united-states-2010-2020.

117 *ruled in Gamble's favor:* Supreme Court, *Estelle v. Gamble,* 429 U.S. 97,

November 30, 1976, https://supreme.justia.com/cases/federal/us/429 /97/.

118 *file a class action lawsuit: Powell v. Ward,* United States Court of Appeals, Second Circuit, September 17, 1976, https://casetext.com/case /powell-v-ward-5/case-summaries.

118 *a right to challenge their solitary confinement: Powell v. Ward,* 1976.

118 *implicated by a judge:* Arnold H. Lubasch, "Prison for Women at Bedford Hills Ordered to Improve Medical Care," *New York Times,* May 1, 1977, https://www.nytimes.com/1977/05/01/archives/prison-for-women -at-bedford-hills-ordered-to-improve-medical-care.html.

120 *penal method of the nineteenth century:* New York Auburn State Prison, *Annual Report for Auburn State Prison*, New York Senate Document, No. 20 (1833).

120 *women shared a single room:* "Prisons: Prisons for Women—History," Law Library: American Law and Legal Information, https://www.law .jrank.org/pages/1799/Prisons-Prisons-Women-History.html.

120 *The ideology of solitary confinement:* Gustave de Beaumont and Alexis de Tocqueville, *On the Penitentiary System in the United States and Its Application in France* (Philadelphia, PA: Carey, Lea, & Blanchard, 1833).

121 *"the reformation of girls ...":* George de Beaumont and de Tocqueville, *On the Penitentiary System in the United States,* 123.

121 *"female convicts were ...":* State of New York, *Senate Documents,* 67th session, vol. 1, no. 20 (1844): 32–33.

121 *"This special disdain ... their male counterparts":* Nicole Hahn Rafter, "Prisons for Women, 1790-1980," *Crime and Justice* 5 (1983): 129–81.

121 *a way to persecute women:* Karlene Faith, *Unruly Women: The Politics and Confinement of Resistance* (Vancouver, Canada: Press Gang Publishers, 1993).

122 *decline in her health:* W. David Lewis, *From Newgate to Dannemora: The Rise of the Penitentiary in New York, 1796–1848* (Ithaca, NY: Cornell University Press, 1965).

123 *DeWitt Clinton objected:* State of New York, Senate Journal, 51st session, 2nd meeting (1828), 103.

123   *authorize the building:* New York Mount Pleasant State Prison, *Report of the Inspectors of the Mount Pleasant State Prison,* New York Senate Documents, no. 20 (1836).

123   *overcrowding in the 1840s:* New York Committee on State Prisons, 1832, *Report on the Committee on State Prisons,* New York Senate Document, No 74 (1832): 9.

123   *provide literacy courses:* Marek Fuchs, "The Women of Sing Sing," *New York Times,* April 21, 2002, https://www.nytimes.com/2002/04 /21/nyregion/the-women-of-sing-sing.html.

125   *looking at the names:* The Women of the ACE Program of the Bedford Hills Correctional Facility, *Breaking the Walls of Silence: AIDS and Women in a New York State Maximum-Security Prison* (New York, New York: The Overlook Press, 1998), 15.

125   *"The sexual transmission ...":* Susan Sontag, *Illness as Metaphor,* 111.

126   *improve the health of inmates:* Close Rikers Campaign, date accessed April 5, 2024, https://www.campaigntocloserikers.org/.

126   *"Generally, there is a quick response ...":* Homer Venters, *Life and Death in Rikers Island* (Baltimore, MD: Johns Hopkins University Press, 2019).

128   *passed a law:* "Private Ownership or Operation of Correctional Facilities," Consolidated laws of New York, Chapter 43, Article 6, Private ownership or operation of correction facilities, August 13, 2021, date accessed April 5, 2024, https://www.nysenate.gov/legislation/laws /COR/121.

129   *a more profound vision of care:* Eric Reinhart, "Medicine for the People," *Boston Review,* March 22, 2021, https://www.bostonreview.net /articles/eric-reinhart-accompaniment-and-medicine/.

129   *"it's possible ... medical profession":* Eric Topol, "Why Doctors Should Organize," *New Yorker,* August 5, 2019, https://www.newyorker.com /culture/annals-of-inquiry/why-doctors-should-organize.

130   *"... striving toward unmitigated totalitarianism":* Angela Y. Davis, *If They Come in the Morning: Voices of Resistance* (London: Verso Books, 2016).

130 *In "An Open Letter to My Sister, Angela Y. Davis":* James Baldwin, "An Open Letter to My Sister, Miss Angela Davis," *New York Review of Books*, January 7, 1971, https://www.nybooks.com/articles/1971/01/07 /an-open-letter-to-my-sister-miss-angela-davis/.

134 *argues scholar Amada Armenta:* Amada Armenta, *Protect, Serve, and Deport: The Rise of Policing as Immigration Enforcement* (Berkeley, CA: University of California Press, 2017).

135 *"...growth of California's state prison system ...":* Ruth Wilson Gilmore, *Golden Gulag: Prisons, Surplus, Crisis, and Opposition in Globalizing* (Berkeley, CA: University of California Press, 2007).

135 *divesting from social programs:* Jackie Wang, *Carceral Capitalism* (Cambridge, MA: MIT Press, 2018).

135 *increased by nearly 500 percent:* Angela Browne, Brenda Miller, and Eugene Maguin, "Prevalence and Severity of Lifetime Physical and Sexual Victimization among Incarcerated Women," *International Journal of Law and Psychiatry* 22, no. 3–4 (May–August 1999): 301–322, https://doi.org/10.1016/S0160-2527(99)00011-4.

136 *A 1999 study at Bedford Hills:* Browne, "Prevalence and Severity of Lifetime Physical and Sexual Victimization."

136 *"reform is not ... punishment":* Maya Schenwar and Victoria Law, *Prison by Any Other Name: The Harmful Consequences of Popular Reforms* (New York: The New Press, 2020).

CHAPTER 5: EBOLA TOWN

140 *"sounds like, smells like Ebola":* "The Doctors: The Ebola Fighters in Their Own Words," *TIME,* December 10, 2014, https://time.com/time -person-of-the-year-ebola-doctors/.

142 *"...the concept of a good death ...":* Josh Mugele and Chad Priest, "A Good Death—Ebola and Sacrifice," *New England Journal of Medicine* 371, no. 13 (September 2014): 1,185–87, https://www.nejm.org/doi /full/10.1056/NEJMp1410301.

137 *expressed Dr. Joanne Liu:* Mark Landler and Somini Sengupta, "Global

Response to Ebola is Too Slow, Obama Warns," *New York Times,* September 25, 2014, https://www.nytimes.com/2014/09/26/world/africa /obama-warns-of-slow-response-to-ebola-crisis.html#:~:text =UNITED%20NATIONS%20—%20Seeking%20to%20speed ,little%20and%20moving%20too%20slowly.

143   *the World Health Organization noted:* World Health Organization, WHO Key Messages on Ebola outbreak in West Africa, October 24, 2014, https://ebolaresponse.un.org/who-key-messages-ebola-outbreak -west-africa.

144   *she rescinded civil liberties:* "Liberia Declares State of Emergency over Ebola Virus," *BBC News*, August 7, 2014, https://www.bbc.com/news /world-28684561.

145   *"August and July . . . symptomatic people":* "The Doctors: The Ebola Fighters."

148   *"We expect the government . . .":* Agence France-Presse, "Ebola Quarantine in Liberia's Capital Sparks Violence in Slum," *The Guardian*, August 21, 2014, https://www.theguardian.com/society/2014/aug/21 /ebola-quarantine-violence-west-point-monrovia-liberia.

148   *Figure:* Amber Hildebrandt, "Ebola Outbreak: Why Liberia's Quarantine in West Point Slum Will Fail," *CBC News,* August 25, 2014, https: //www.cbc.ca/news/world/ebola-outbreak-why-liberia-s-quarantine -in-west-point-slum-will-fail-1.2744292.

149   *Liberian workers without a job:* World Bank, "Nearly Half of Liberia's Workforce No Longer Working since Start of Ebola Crisis," November 19, 2014, https://www.worldbank.org/en/news/press-release/2014/11 /19/half-liberia-workforce-no-longer-working-ebola-crisis.

149   *". . . nowhere to go for our daily bread":* Norimitsu Onishi, "Clashes Erupt as Liberia Sets an Ebola Quarantine," *New York Times,* August 20, 2014, https://www.nytimes.com/2014/08/21/world/africa /ebola-outbreak-liberia-quarantine.html.

150   *"It is inhumane . . .":* Agence France-Presse, "Ebola Quarantine."

150   *"You fight Ebola with arms?":* Onishi, "Clashes Erupt."

151   *"assessment about the value of life":* Adia Benton, "Race and the

Immuno-Logics of Ebola Response in West Africa," *Somatosphere*, September 19, 2014, http://somatosphere.net/2014/race-and-the -immuno-logics-of-ebola-response-in-west-africa.

151 *"The government . . . now empty":* Clair MacDougall, "Liberian Government's Blunders Pile Up in the Grip of Ebola," *TIME*, September 2, 2014, https://time.com/3247089/liberia-west-point-quarantine-monro via/.

152 *"While decisive action is needed . . .":* Amesh Adalja, "Quarantining an entire Liberian Slum to Fight Ebola Is a Recipe for Disaster," *Washington Post*, August 28, 2014, https://www.washingtonpost.com/news /to-your-health/wp/2014/08/28/quarantining-an-entire-liberian -slum-to-fight-ebola-is-a-recipe-for-disaster/.

152 *"slow death":* Lauren Berlant, "Slow Death (Sovereignty, Obesity, Lateral Agency)," *Critical Inquiry* 33 no. 4 (Summer 2007): 754–80.

153 *"a nasty new contagious disease . . .":* Peter Piot, *No Time to Lose: A Life in Pursuit of Deadly Viruses* (New York: W. W. Norton & Company, 2012).

153 *noted the self-imposed measures:* Piot, *No Time to Lose.*

154 *thought the outbreak:* Mark Honigsbaum, "Jean-Jacques Muyembe-Tamfum: Africa's Veteran Ebola hunter," *The Lancet* 385 no. 9986 (June 20, 2015): 2,455. https://doi.org/10.1016/S0140-6736(15)61 128-X.

154 *he collected blood samples:* Honigsbaum, "Jean-Jacques Muyembe-Tamfum."

155 *"Movements for animal rights . . .":* Donna Haraway, "A Manifesto for Cyborgs: Science, Technology, and Socialist Feminism in the 1980s," *Australian Feminist Studies* 2 no. 4 (Autumn 1987): 1–42, https://doi .org/10.1080/08164649.1987.9961538.

156 *the genome of nearly thirty strains:* Nicholas Di Paola et al., "Viral Genomics in Ebola Virus Research," *Nature Reviews Microbiology* 18 no. 7 (May 4, 2020): 365–78, https://doi.org/10.1038/s41579-020-03 54-7.

158 *"income generated by foreign concessions . . .":* Stephen Ellis, *The Mask*

of Anarchy: The Destruction of Liberia and the Religious Roots of an African Civil War (London: Hurst & Co Ltd, 2001), 47.

159 *spent $100.5 million on healthcare:* Liberia Health Expenditure, *The World Bank,* accessed April 2, 2024, https://data.worldbank.org /indicator/SH.XPD.CHEX.GD.ZS?locations=LR.

160 *"Everybody knows ... by surprise":* Albert Camus, *The Plague* (New York: The Modern Library, 1948).

160 *assisted these organizations:* Krista C. Swanson et al., "Contact Tracing Performance during the Ebola Epidemic in Liberia, 2014–2015," *PLoS Neglected Tropical Diseases* 12 no. 9 (September 2018): e0006762, https://doi.org/10.1371/journal.pntd.0006762.

160 *the procedure can be limiting:* Benjamin Armbruster and Margaret L. Brandeau, "Contact Tracing to Control Infectious Disease: When Enough Is Enough," *Health Care Management Science* 10 no. 4 (December 2007): 341–55, https://doi.org/10.1007/s10729-007-9027-6.

161 *"ultimately aims to reduce ...":* Saurabh Shrivastava and Prateek Shrivastava, "Role of Contact Tracing in Containing the 2014 Ebola Outbreak: A Review," *African Health Sciences* 17 no. 1 (March 2017): 225–36, https://www.ajol.info/index.php/ahs/article/view/156380.

161 *"The human causes . . . latex gloves":* Mike McGovern, "Bushmeat and the Politics of Disgust," *Cultural Anthropology,* October 7, 2014, https://culanth.org/fieldsights/bushmeat-and-the-politics-of-disgust.

161 *a compound of forces:* Adia Benton and Kim Yi Dionne, "5 Things You Should Read before Saying the IMF is Blameless in the 2014 Ebola Outbreak," *Washington Post,* January 5, 2015, https://www.washingtonpost .com/news/monkey-cage/wp/2015/01/05/5-things-you-should-read -before-saying-the-imf-is-blameless-in-the-2014-ebola-outbreak/.

162 *"the relationship between public health and medicine . . .":* Allan M. Brandt and Martha Gardner, "Antagonism and Accommodation: Interpreting the Relationship Between Public Health and Medicine in the United States During the 20th Century," *American Journal of Public Health* 90 no. 5 (May 2000): 707–715, https://www.ncbi.nlm.nih .gov/pmc/articles/PMC1446218/pdf/10800418.pdf.

162 *"Ebola was not simply . . .":* Ibrahim Abdullah and Ismail Rashid, *Understanding West Africa's Ebola Epidemic: Towards a Political Economy* (London, Zed Books, 2017).

162 *"Old disease in a new context...":* Kevin Sack et al., "How Ebola Roared Back," *New York Times,* December 29, 2014, https://www.nytimes.com /2014/12/30/health/how-ebola-roared-back.html.

163 *60 percent of respondents:* Y Care International, "The Ebola Outbreak in Liberia: Young People's Needs in the West Point Slum," October 2, 2014, accessed April 2, 2024, https://reliefweb.int/report/liberia/ebola -outbreak-liberia-young-people-s-needs-west-point-slum.

163 *a health worker noted:* Y International, "The Ebola Outbreak in Liberia."

163 *"by seizing their cadavers . . .":* Honigsbaum, "Jean-Jacques Muyembe-Tamfum."

165 *describe the stakes of quarantine:* Geoff Manaugh and Nicola Twilley, *Until Proven Safe: The Gripping History of Quarantine, from the Black Death to the Post-Covid Future* (London: Picador, 2021), 16.

166 *"Among the patients with EVD . . .":* Timothy M. Uyeki, et al., "Clinical Management of Ebola Virus Disease in the United States and Europe," *New England Journal of Medicine* 374 no. 7 (2016): 636–46, https://www.nejm.org/doi/full/10.1056/NEJMoa1504874.

166 *stoking fear:* Adam Nossiter, "Fear of Ebola Breeds a Terror of Physicians," *New York Times,* July 27, 2014, https://www.nytimes.com/2014 /07/28/world/africa/ebola-epidemic-west-africa-guinea.html.

166 *"it becomes painfully apparent . . .":* Reena Pattani, "Unsanctioned Travel Restrictions Related to Ebola Unravel the Global Social Contract," *Canadian Medical Association Journal* 187 no. 3 (2015): 166–67, https://doi.org/10.1503/cmaj.141488.

167 *"expresses pity but also conveys contempt":* Sontag, *Illness as Metaphor,* 49.

168 *"humans can carry the microbe . . .":* Sonia Shah, *Pandemic: Tracking Contagions, from Cholera to Ebola and Beyond* (New York: Farrar, Straus and Giroux, 2017), 39.

171 *begins with a question:* Kaj Larsen, "Monkey Meat and the Ebola Outbreak in Liberia," VICE News, YouTube, June 26, 2014, date accessed April 2, 2024, https://www.youtube.com/watch?v=XasTcDsDfMg.

172 *"...wild meat without incident":* Jesse Bonwitt et al., "Unintended Consequences of the 'Bushmeat Ban' in West Africa during the 2013–2016 Ebola Virus Disease Epidemic," *Social Science & Medicine* 200 (March 2018): 166–173, https://doi.org/10.1016/j.socscimed.2017.12.028.

174 *conceptualization of "disgust":* William Ian Miller, *The Anatomy of Disgust* (Cambridge, MA: Harvard University Press, 1998).

174 *"To photograph is ... like power":* Susan Sontag, *On Photography* (New York: Picador, 1973).

174 *fueling American fears:* Gerard Flynn and Susan Scutti, "Smuggled Bushmeat Is Ebola's Back Door to America," *Newsweek Magazine,* August 21, 2014, https://www.newsweek.com/2014/08/29/smuggled-bushmeat-ebolas-back-door-america-265668.html.

175 *Figure:* Patrick Chappatte, "Dealing with Ebola," *New York Times,* August 26, 2014, https://www.nytimes.com/2014/08/26/opinion/patrick-chappatte-the-ebola-epidemic.html.

179 *"That means ... drugs like statins":* James Surowiecki, "Ebolanomics," *New Yorker*, August 18, 2014, https://www.newyorker.com/magazine/2014/08/25/ebolanomics.

180 *"...malaria-free into malaria zones":* Emmanuel Akyeampong, "Disease in West African History," in *Themes in West Africa's History,* ed. Emmanuel Akyeampong (Suffolk, UK: Boydell & Brewer, 2006), 186–207.

180 *In his 1938 essay:* Bertolt Brecht, "The World's One Hope," 1938.

181 *"to address the health problems ...":* Philip Ireland, "Ebola: What My Experience Taught Me," *CNN,* July 9, 2015, https://edition.cnn.com/2015/07/09/opinions/ireland-ebola-response/index.html.

CHAPTER 6: RELENTLESS

183 *"Grief, then ...":* Claudia Rankine, "The Condition of Black Life Is One of Mourning," *New York Times,* June 22, 2015, https://www

.nytimes.com/2015/06/22/magazine/the-condition-of-black-life
-is-one-of-mourning.html?_r=0.

189 Guardian *health editor:* Andrew Gregory, "UK Government Accused
of 'Explaining Away' Covid Race Issues," *The Guardian,* May 5, 2022,
https://www.theguardian.com/uk-news/2022/may/05/uk-government
-accused-of-explaining-away-covid-race-issues.

200 *"To become a writer . . .":* Deborah Levy, *Things I Don't Want to Know*
(New York: Penguin Random House, 2018).

201 *equated with inordinate stamina:* Michele Wallace, *Black Macho and
the Myth of the Superwoman* (London: Verso Books, 1999).

202 *". . . When, if ever, will we have reckoned?":* Yaa Gyasi, "White People,
Black Authors Are Not Your Medicine," *The Guardian,* March 20,
2021, https://www.theguardian.com/books/2021/mar/20/white-people
-black-authors-are-not-your-medicine.

203 *Germany is riddled:* Katrin Bennhold, "QAnon Is Thriving in Germany.
The Extreme Right Is Delighted," *New York Times,* October 11, 2020,
https://www.nytimes.com/2020/10/11/world/europe/qanon-is-thriving
-in-germany-the-extreme-right-is-delighted.html.

212 *"When we love . . .":* Paulo Coelho, *The Alchemist* (New York: Harper
One, 2014).

214 *captures the city's pulse:* Claude McKay, *Romance in Marseille* (New
York: Penguin Random House, 2020).

215 *"this macabre roll call . . .":* Keeanga-Yamahtta Taylor, "The Black
Plague," *New Yorker,* April 16, 2020, https://www.newyorker.com/news
/our-columnists/the-black-plague.

216 *"For whom and at what cost":* Donna Jeanne Haraway, *Primate Visions:
Gender, Race, and Nature in the World of Modern Science* (New York:
Routledge, 1989), 1.

CHAPTER 7: LOCKED UP

219 *"Of course . . . the alphabet":* Octavia E. Butler, *A Few Rules for Predicting
the Future: An Essay* (San Francisco, CA: Chronicle Books, 2024).

220 *the conditions that led to:* "Reyalite prizonye yo anndan Prizon Pòto-prens. Plizyè Depòte (USA) denonse kondisyon otorite yo Chwazi lage yo anndan prizon akoz daprè," Facebook video, June 30, 2022, https://www.facebook.com/theriel.thelus/videos/reyalite-prizonye-yo-anndan-prizon-pòtoprens-plizyè-depòte-usa-denonse-kondisyon/1512340692558604/.

221 *81 percent of incarcerated people:* World Prison Brief Data, date accessed April 5, 2024, https://www.prisonstudies.org/country/haiti.

221 *"the illegal and arbitrary use . . .":* "'N ap mouri': Report on Conditions of Detention in Haiti," United Nations Human Rights Office of the High Commissioner, June 25, 2021, https://www.ohchr.org/sites/default/files/Documents/Countries/HT/2021-06-25-executive-summary-en.pdf.

221 *could not afford to pay:* Jonathan Lippman, "A Test of What Society Considers Morally Unacceptable," *New York Times,* November 10, 2021, https://www.nytimes.com/2021/11/10/opinion/rikers-island-jail.html.

221 *". . . designed for 800 inmates housed 4,000 people":* Jacqueline Charles, "U.N. Secretary-General proposes rapid-response troops to help Haiti regain control from gangs," *Miami Herald,* October 10, 2022, https://www.miamiherald.com/news/nation-world/world/americas/haiti/article267082601.html.

221 *"What does it mean to defend the dead?":* Christina Sharpe, *In the Wake: On Blackness and Being* (Durham, NC: Duke University Press, 2016), 10.

222 *infected over 800,000 Haitians:* Elizabeth C. Lee, et al., "Achieving Coordinated National Immunity and Cholera Elimination in Haiti through Vaccination: A Modelling Study," *The Lancet Global Health* 8 no. 8 (August 2020): e1081–89, https://doi.org/10.1016/S2214-109X(20)30310-7.

222 *"To die from cholera . . .":* Loune Viaud et al., "A Cholera Outbreak in a Haitian Prison Threatens to Kill Hundreds Within Days," *The Nation,* October 11, 2022, https://www.thenation.com/article/world/haiti-prison-cholera-outbreak/.

223 *the men he had been imprisoned with:* Tanvi Misra, "The Pipeline Funneling US Deportees to Haitian Prison," *The Nation,* November 14, 2022, https://www.thenation.com/article/society/deportation -haiti-prison-conditions/.

224 *"The Dominican Republic persistently deports ...":* Haydi Torres and Eleni Baskt, "Haitian Deportees Face Wretched, Indefinite, Illegal Detention in Haiti," *Haiti Liberté,* August 17, 2022, https://haitiliberte.com /haitian-deportees-face-wretched-indefinite-illegal-detention-in-haiti/.

225 *twenty million motorists stopped by police:* The Stanford Open Policing Project, date accessed, April 5, 2024, https://openpolicing.stanford .edu/findings/.

225 *one of at least thirty Haitian men:* "Haitians Being Returned to a Country in Chaos," *Human Rights Watch,* March 24, 2022, https://www.hrw .org/news/2022/03/24/haitians-being-returned-country-chaos.

225 *Some were deported after:* Eileen Sullivan, "U.S. Accelerated Expulsions of Haitian Migrants in May," *New York Times,* June 9, 2022, https: //www.nytimes.com/2022/06/09/us/politics/haiti-migrants-biden .html.

226 *psychological and physical torment:* Homer Venters, *Life and Death in Rikers Island* (Baltimore, MD: Johns Hopkins University Press, 2019).

227 *Roody Fogg:* Hannah Adely, "They Threw Him in a Corner: After Cholera Outbreak, One Deportee Is Dead, Two Others Ill in Haiti Prison," *North Jersey,* October 15, 2022, https://eu.northjersey.com/story/news /2022/10/15/haitian-deportees-dying-after-cholera-outbreak -in-prison/69559393007/#:~:text=Roody%20Fogg%2C%20 40%2C%20died%20on,said%20in%20an%20emailed%20statement.

227 *relatives in the United States and Haiti:* Leonardo March, "Families, Activists Demand Release of Haitians Deported after Serving Time in US," *Haitian Times,* August 25, 2022, https://haitiantimes.com/2022/08 /25/photos-families-activists-demand-release-of-haitians-deported -after-serving-time-in-us/.

227 *"caught in [a] crossfire ..."* Tanvi Misra, "The Pipeline Funneling US

Deportees to Haitian Prison," *The Nation,* November 14, 2022, https://www.thenation.com/article/society/deportation-haiti-prison-conditions/.

228 *"I don't feel right…":* Luke Taylor, "They Have No Fear and Mercy: Gang Rule Engulfs Haitian Capital," *The Guardian,* September 18, 2022, https://www.theguardian.com/world/2022/sep/18/haiti-violence-gang-rule-port-au-prince.

228 *"… fight for your life as though it were our own …":* James Baldwin, "An Open Letter to My Sister, Miss Angela Davis," *New York Review of Books,* January 7, 1971, https://www.nybooks.com/articles/1971/01/07/an-open-letter-to-my-sister-miss-angela-davis/.

POSTSCRIPT

231 *"Epidemics … contamination of foreigners":* Sontag, *Illness as Metaphor,* 166.

232 *"… unilaterally, decisively, with few checks on his power …":* Masha Gessen, *Surviving Autocracy* (New York: Riverhead Books, 2000), 230.

233 *The United States exceeded over 1.2 million:* "United States—COVID-19 Overview—Johns Hopkins," Johns Hopkins Coronavirus Resource Center, accessed February 16, 2024, https://coronavirus.jhu.edu/region/united-states.

234 *"undeserving" and "deserving":* Beatrice Adler-Bolton and Artie Vierkant, *Health Communism* (London: Verso Books, 2022).

Index

A

Abdullah, Ibrahim, 162
abortion, 26
ACE (AIDS Committee for
    Education), 109–16, 118–119,
    124, 131, 132, 136
Action Contre la Faim, 160
ACT UP, 112–13
Adalja, Amesh, 152
Africa
    colonialism in, 62, 72, 156, 161
    Anglo-Boer War and, 54
    Berlin Conference ("Scramble
        for Africa") and, 157
    see also Africa, German colonies
        in
    monkeys in, 171–75
    West Africa, see West Africa
Africa, German colonies in, 47,
    52–56, 57, 66, 68–70, 72
    as medical laboratory, 39–72
    concentration camps in, xxviii,
        44–45, 54–67
    race science and, 67–69
    for sleeping sickness research,
        40–49, 52–66, 71

    political concentration camps in,
        54–55, 57, 68
Africa as a Living Laboratory
    (Tilley), 47
Africa CDC, 178
African Americans
    believed to be biologically
        distinct from white people,
        4–8, 19–20, 70
    Civil Rights movement and, 106,
        110, 119, 131, 201, 202, 217
    Covid and, 206–7, 215, 217
    doctors, 33–36
    elected to government, 31
    formerly enslaved, 31–35
    Liberia as home for, 156–57
    health care for, 32, 33, 35
    health disparities between white
        people and, 35
    medical students, 33–36
    self-hatred in, 131
    slaves, see slaves
    violence against, 202
    women, 6, 201
African Journal of Health Sciences,
    160–61

Afro-Caribbean people, xx, 156, 185

see also Haitian immigrants

AIDS, *see* HIV/AIDS

"AIDS and Its Metaphors" (Sontag), 231

AIDS Coalition to Unleash Power, xxviii

AIDS Committee for Education (ACE), 109–16, 118–119, 124, 131, 132, 136

Akyeampong, Emmanuel, 179–80

Alchemist, The (Coelho), 212

Alert to AIDS, 113

Alexander, Tsar, 79

Alter, Alexander, 97

Alternative für Deutschland (AfD), 194

American Colonization Society (ACS), 156

Anan, David, 150

Ananthavinayagan, Thamil, 71

Anglo-Boer War, 54

anthrax, 49, 50

antibiotics, xiv, xxviii

anxiety, xv, 84

Aristotle, 61, 62

Armenta, Amada, 134–35

arsenic, 4

arsenophenylglycin, 64–65

atoxyl, 45, 60–62

Artibonite River, 222

asylum seekers and refugees

in Germany, 184, 192–200

in Greece, 196

US-Mexico border and, 232

atoxyl, 45, 60–62

Attica prison riot, 117

Auburn State Prison, 119–23, 135

August Rebellion, 117

Aunt N, 133–34

avian flu, 143

Aviv, Rachel, 84

B

bacteria, 4, 42, 50

see also microbes, microbiology

Baldwin, James, 96, 107, 131, 217, 228

Bauche, Manuela, 59–60

Beaumont, Gustave de, 120–121

Beck, Max, 41

Bedard, Rachael, 126–29

Bedford Hills, NY, 108–9

Bedford Hills Correctional Facility, 108–9, 116–17, 129, 136

AIDS Committee for Education (ACE) at, 109–16, 118–119, 124, 131, 132, 136

AIDS quilt at, 124–125, 132

August Rebellion at, 117

Boudin in, 105–16, 119, 135, 136

children of women in, 105–6, 111

HIV/AIDS at, 106, 108–15, 116, 119, 125, 131–32
writing by inmates of, 130–32
Bedford Hills State Reformatory, 124–25
Bellevue Penitentiary, 123
Bennhold, Katrin, 203
Benton, Adia, 151, 161, 169
Berche, Patrick, 79
Berlant, Lauren, 152
Berlin, xxvi, 42, 44, 45, 50, 51, 67, 69, 70, 79, 186, 191–93, 195, 196, 198, 199, 203, 204, 206–9, 211, 217
Berlin Conference ("Scramble for Africa"), 157
Bernabe, Saguens, 223–28
Biden, Joseph, 232
Black Americans, see African Americans
Black Lives Matter, 70
Black Macho and the Myth of the Superwoman (Wallace), 201
Black Panther Party, 234
"Black Plague, The" (Taylor), 215
bloodletting, 5–6, 8, 19
Bloomsbury group, 88
bodily autonomy, xxvii
Boers, 54
Boggs, Grace Lee, 100
Boudin, Kathy, 105–8
    in armored truck robbery, 106–107

at Bedford Hills Correctional Facility, 105–16, 119, 135, 136
HIV-positive inmates and, 106, 108–14, 116, 118
in Weather Underground, 106–7
Boudin, Leonard, 105
Boston Psychiatric Hospital, 83
Boston University, 34
Bowen, James L., 34
Boyer, Anne, xi, 99
Brandt, Allan, 162
Breaking the Walls of Silence (women of the ACE program), 131
Brecht, Bertolt, 180
Brisbane, Samuel, 141–42
Britain, 152, 191, 206
    in Anglo-Boer War, 54
    Covid in, 188–89, 199
    Defence of the Realm Act in, 80–81
    influenza pandemic in, 79–85, 89, 91, 98
    racism in, 188–89
British Royal Society, 46
British Society for Sleeping Sickness, 56
Brittin, William, 119–120
Broad Street Pump, 16–17
Brooks, George W., 34
bubonic plague, 39, 85
    quarantine for, 165

Bugala research camp, 57–59

Bulletin of Woman's Medical College of Pennsylvania, 76

bushmeat, 171–75

Butler, Octavia, 219

C

Cabin, Quarter, Plantation (Ellis and Ginsburg), 7

Cabral, Amílcar, 183

Calanques, 213, 214

California, 135

calomel, 14

Camp Funston, 73–76, 86

Camus, Albert, 159–60

cancer, 96–100, 142, 197, 204, 205

chemotherapy for, 201–2

Covid-19 and, 184, 200–205

capitalism, xxii, 144, 233

Carceral Capitalism (Wang), 135

Cartwright, Samuel A., 1–7

Cassis, 213–15

castor oil, 8, 14, 19, 21

Castro, Fidel, 105

Centers for Disease Control and Prevention (CDC), xvii, 109, 160

Chan, Margaret, 162

Chappatte, Patrick, 175–76

China, 143, 206

Choctaw people, 17–18

cholera, xxviii, 11, 12, 14–15, 19, 32, 39, 49, 68, 79, 140, 168

in Haiti, 221–23, 227, 229

research on, 15–17

slaves and, 3–7, 9–15, 17, 19, 21

theories about cause of, 4–5, 16, 19

water and, 12, 16–17

Civil Rights movement, 106, 110, 119, 131, 201, 202, 217

Civil War, 1, 9, 11, 12, 31, 32, 35

Clark, Judith, 109–11, 114, 118, 136

Clinton, DeWitt, 123

Cobb, Ebenezer B., 122, 123

Coelho, Paulo, 212

colonialism, xviii, xxii, xxix, 7, 161

in Africa, 62, 72, 156, 161

Anglo-Boer War and, 54

Berlin Conference ("Scramble for Africa") and, 157

see also Africa, German colonies in

concentration camps, 45–46, 54–55

in Germany's African colonies, 54–55, 57, 68

medical, xxviii, 45, 54–67

political, 54–55, 57, 68

in Nazi Germany, 45, 54

"Condition of Black Life Is One of Mourning, The" (Rankine), 183–84

Confederate artifacts, 70

confinement, xv, xxiii–xxv, xxvii–xxviii, 103, 233–34

in author's bout with typhoid fever, xi–xv, xxvi, xxvii, xxix
in concentration camps, *see* concentration camps
in Ebola isolation centers, 146–47
in Haitian community, xviii, xix, xxiii, xxvi
lockdowns, 151, 233
Covid, 147, 187–89, 192–94, 196–98, 202, 205, 206, 218, 232
Ebola, in West Point, Liberia, xxvii–xxviii, 146–53, 168–70, 176–81
of migrants, xxiii, 133–35
quarantine, 152, 164–65, 168, 233
for bubonic plague, 165
history of, 165
for influenza, 77, 82, 84
for sleeping sickness, 61, 63–64
to sickbed, 75, 86, 87, 91, 97, 99–100
in slavery, 3–5, 7, 12, 26–27, 38
Woolf and, 87, 102
Congo, 153–54
constitutional amendments
Eighth, in *Estelle v. Gamble*, 117, 129
Fourteenth, 118
Cooper, Isaac, 21
Copenhagen, 79, 207–11

Corpus Hippocraticum (Hippocrates), 78
Covid-19 pandemic, xxii, xxiv–xxix, 69–71, 183–218, 220, 232
and asylum seekers in United States, 232
author's marriage during, 207–12
Black Americans and, 206–7, 215, 217
in Britain, 188–89, 199
conspiracy theories and, xxvi, xxvii, 203, 218
Covid Zero policies for, xxvi
deaths related to, 233
in Germany, 69, 183, 187, 203
asylum seekers and, 184, 192–200
cancer patients and, 184, 200–205
lockdown in, 187–89, 192–94, 196–98, 202, 205, 206, 218
sex workers and, 184–92
vaccines and, 203–5
healthcare workers and, 142
literature and, 97
lockdowns in, 147, 187–89, 192–94, 196–98, 202, 205, 206, 218, 232
prisons and, 220, 226–27
Trump and, 206
vaccines and, xxiv, 71, 198–99, 203–5

Criterion, 92
crocodile, 41
Crooks, Carol, 117
Crumpler, Rebecca, 34, 35
Curie, Marie, 39, 40

D
Danticat, Edwidge, 229
Davis, Angela, 26, 131, 228
Dawn of Everything, The (Graeber),
    132
Dear Science (McKittrick), xviii–
    xix
Defence of the Realm Act, 80–81
Democratic Republic of Congo,
    153–54
Denmark, 207–8, 211
depression, 84
Dickinson, Violet, 89
Dionne, Kim Yi, 161
Discipline and Punish (Foucault),
    105, 114–115
disgust, 174
doctors, *see* physicians
Doctors Without Borders, 142,
    146
Dominican Republic, 224
"do no harm," 30
Douglass, Frederick, 20–21
Drake, Lee W., 73, 74
drug overdoses, xxii
Dr. Wise on Influenza, 82
Du Bois, W. E. B., 31, 35, 139

due process, 118
Duvalier, Jean-Claude, xvi, xviii
Dying Colonialism, A (Fanon), 139
dysentery, 66

E
Ebola
    discovery of virus, 154
    genome of, 156
    1976 outbreak of, 153–54
    origins of, 155
    research on, 154–56
    risk of death from, 166–67,
        170
    symptoms of, 139, 140, 143
    treatment for, 166, 179
    vaccines for, 179
    in West Africa in 2014, 143–44,
        146, 152, 164, 166, 176, 178
    travel restrictions and, 151,
        165–68, 170–71, 176
Ebola outbreak in Liberia (2014),
    xxviii, 139–81
    cartoons about, 175–76
    contact with the sick and the
        dead during, 163–64
    death toll from, 178
    end of, 178
    healthcare system and, 143, 161,
        170
    healthcare workers and, 139–45,
        160–63, 180–81
    isolation centers in, 146–47

media representation of,
171–76

militarization in, xxvii, 148–52,
168–69

monkey meat and, 171–75

public health measures in,
144–53, 177, 180

skepticism about, 163, 164,
171–73

tracking and tracing of, 160–61

travel restrictions and, 151,
165–68, 170–71, 176

Vice Media video on, 171–75

West Point lockdown in,
xxvii–xxviii, 146–53, 168–70, 176–81

Eckart, Wolfgang, 53

Ehrlich, Paul, 64–65

Eighth Amendment, *Estelle v. Gamble* and, 117, 129

Einstein, Albert, 50

electric lighting, 80

Eliot, T. S., 92

Elkeles, Barbara, 62

Ellis, Clifton, 7

Ellis, Stephen, 157–58

Emre, Merve, 96

endometriosis, 197

Entebbe, 44

Epsom salts, 21

Estelle, W. J., Jr., 117, 129

Ethiopia, 62

Ewell, James, 19

F

Fair, Joseph, 171–73

Faith, Karlene, 121

Fallah, Mosoka, 145

Farmer, Paul, 164

Fanon, Frantz, 139

Feldman, Oskar, 44, 52

firearms, 224

Firestone Tire Company, 158

First World War, 73, 74, 76, 77, 80,
81, 84, 85, 91, 93, 94

Fischer, Eugen, 67, 68

Fiske, A. S., 32

Flexner, Abraham, 35–36

Florida, 135–36, 217

Floyd, George, 69, 137, 202, 217

flu, *see* influenza

Fogg, Roody, 227

Fort Devens, 76

Fort Riley, KS, 73–74

Foucault, Michel, 105, 114–115

Fourteenth Amendment, 118

Francis, Thomas, 101

Freedmen's Bureau, 31–35

Freedmen's Hospital, 33

freedom, xxiv, xxvii, xxviii, 28, 218

Freedom (Patterson), xxiv

Freire, Paulo, 107

Freud, Sigmund, xv

Fugitive Slave Act, 30

G

Gamble, J. W., 117, 129
Gardner, Martha, 162
genotype and phenotype, 155
Georgia, 9
Germany, 49, 54, 207
  African colonies of, *see* Africa,
    German colonies in
  Berlin, xxvi, 42, 44, 45, 50, 51, 67,
    69, 70, 79, 186, 191–93, 195,
    196, 198, 200, 203, 204, 206–9,
    211, 217
  Black life in, 184–85, 200–202,
    204
  Covid-19 pandemic in, 69, 183,
    187, 203
  asylum seekers and, 184,
    192–200
  cancer patients and, 184,
    200–205
  lockdown in, 187–89, 192–94,
    196–98, 202, 205, 206, 218
  sex workers and, 184–92
  vaccines and, 203–5
  nationalism in, 47
  Nazi, 66–68, 72
  concentration camps in, 45, 54
  parliamentary elections in, 194
  racism in, 202
germ theory, 50, 56–57, 79, 84
Gessen, Masha, 232
Gilbert, David, 106
Gilmore, Ruth Wilson, 135

Ginsburg, Rebecca, 7
Gitchell, Albert, 73, 74
Golden Gulag (Gilmore), 135
Gowrie Plantation, 9–15, 17
Graeber, David, 132
Gramsci, Antonio, 115–16
Grand Cape Mount, 148
Grand Junction Waterworks, 16
Greece, 196
Greece, ancient, 78
Green, Sam, 106
Gregory, Andrew, 189
Greville Memoirs, The (Greville), 87
Guardian, 189
Guinea, 151, 156, 165–68, 178
Gyasi, Yaa, 202

H

Haiti, xv–xvi, xviii, 213, 219–29
  cholera epidemic in, 221–23,
    227, 229
  earthquake in, 221–22
  prisoners in, 219–21, 223,
    225–29, 231
  Revolution in, xviii
Haiti (Trouillot), xvi
Haitian Creole, xvi, xviii–xx, xxiii,
  219, 223
Haitian immigrants, xvi–xx, 224
  deportations of, 224, 225
  HIV/AIDS and, xvii, xix
  in Little Haiti, xvii–xx, xxiii, xxvi,
    134

Haitian National Penitentiary, 219–21, 223, 225–29

Haitian Times, 227

Haraway, Donna, 155, 216

Harris, Stephen, 6

Hartman, Saidiya, 1, 17, 25

Haslip, Katrina, 109–10, 114, 136

health, xxi, xxii

"Health and Physique of the Negro American, The" (Du Bois), 35

health care, xxii, xxiv–xxvi
   for African Americans, 32, 33, 35
   Medicare for All, 233
   in United States, 233–34
   see also physicians

Herero, 68

Hippocrates, 30, 78

Hitler, Adolf, 68

HIV/AIDS, xxviii, 231
   ACT UP and, 112–13
   Haitians and, xvii, xix
   high risk groups for, xvii
   judgment about, 125–26
   in New York, 108, 126
   prisons and, xxviii, 106, 108–15, 116, 119, 125, 127, 132–33, 137
   quilt and, 125, 132

Hogarth, Rana, 8

Hogarth Press, 92

Holocaust, 45–46, 54–55

home, access to, 100

Horniblow, Delilah, 22, 23

Horniblow, Margaret, 22

Horniblow, Molly, 28

Howard University, 33–34, 36

How Europe Underdeveloped Africa (Rodney), 62

Huggins, Ericka, 105

Human Development Index, 159

Huntington Unit prison, 116

Hurby, Adolph, 73, 74

Hurston, Zora Neale, 201

hysterectomy, 197

I

If They Come in the Morning (Davis), 130

illness, xi, xxi–xxiii, xxvii, xxviii, 180
   environment as cause of, 4–5, 50
   perceptions of, 96–97, 126

Illness as Metaphor (Sontag), 96–97

immigrants, 134
   Afro-Caribbean, xx
   asylum seekers, 232
   detention centers for, xxiii, 133–35
   Haitian, xvi–xx, 224
   deportations of, 224, 225
   HIV/AIDS and, xvii, xix
   in Little Haiti, xvii–xx, xxiii, xxvi, 134

Incidents in the Life of a Slave Girl (Jacobs), 22

Indian Removal Act, 18

Inflamed (Marya and Patel), xxi–xxii

influenza, xxviii, 77–85
  deaths from, 101
  early accounts of, 78–79
  genome of, 101
  as metaphor for war, 78
  vaccine for, 101
  waves of, 79

influenza pandemic of 1918–1919, 73–103
  bed access and, 100–101
  in Britain, 79–84, 89, 91, 98
  Camp Funston in, 73–76, 86
  creative class and, 85–98
  death toll from, 77–78, 84
  electric lighting blamed for, 80
  Fort Devens in, 76
  literature and, 95–96
  Look Homeward, Angel, 95–96
  Mrs. Dalloway, 91–96
  Pale Horse, Pale Rider, 95
  physicians and, 76, 79
  psychosis and, 83–84
  public health practices and, 77, 81–82, 84
  quarantine in, 77, 82, 84
  in Russia, 79, 81
  socioeconomic effects of, 76
  as "Spanish flu," 81
  Woolf and, 85–99, 101–3

World War I and, 73, 74, 76, 77, 80, 81, 84, 85, 91, 93, 94

injuries, xxii, 7

International Sanitary Conference, 166

ipecac, 6

Ireland, Philip, 139–41, 143, 181

Italy, 62, 115, 152

ivermectin, 8

J

Jackson Memorial Hospital, xi–xv

Jacobs, Harriet, 22–31, 37–38

Jefferson, Thomas, 5–6, 19

John F. Kennedy Memorial Medical Center, 141, 143

Johnson, Boris, 188–89

Journal of Tropical Medicine, 67

Julney, Patrick, 223

K

Kahlo, Frida, 100

Kaiser Wilhelm Institute of Anthropology, 67

Kennedy, Jonathan, 199

Kenya, 196–97, 199

Keynes, Maynard, 88

Kigarama, 64

Killing the Black Body (Roberts), xx–xxi, 27–28

Kleine, Friedrich Karl, 40–42, 63, 64

Knox, Elijah, 22, 23

Koch, Hedwig, 41, 63

Koch, Robert, 39, 49, 70
  legacy of, 70–72
  lymphangitis contracted by, 63
  microbiology career of, 48–52, 56, 62–63, 68
  postulates in, 51
  Nobel Prize awarded to, 51–52
  on scientific career, 49
  sleeping sickness research of, 40–49, 52–66, 71
  arsenic-based treatments in, 45, 60–62
  medical concentration camps in, 45, 54–66

Kountché, Seyni, 200

Krome North Service Processing Center, 133–35

L

Lake Tanganyika, 64

Lake Victoria, 42–44, 52, 53, 56, 64

Land-Lease, 157

Larsen, Kaj, 171–74

laudanum, 19

Law, Victoria, 118, 136

Lazard, Carolyn, 99

Lemon City, Miami (Little Haiti), xvii–xx, xxiii, xxvi, 134

Levy, Deborah, 200

Libbertz, Arnold, 41

Liberia, 156–59, 170, 180
  American policies and, 157–58, 170
  Civil War in, 148, 158, 159, 169, 170
  Ebola in, *see* Ebola outbreak in Liberia
  healthcare system in, 159, 161, 170, 177, 178, 181
  as home for formerly enslaved people, 156–57
  in Human Development Index, 159
  hut tax in, 158
  independence of, 157
  inequality in, 162
  Monrovia, 141, 144–46, 171, 173
  National Public Health Institute in, 177, 178
  Truth and Reconciliation Commission in, 159

Liberian Institute for Biomedical Research, 171

Life and Death in Rikers Island (Venters), 126, 226

life expectancy, xxii, 159

literature, illness in, 92, 97–98, 100
  cancer, 96–99
  Covid pandemic, 97
  influenza pandemic, 95–96
  Look Homeward, Angel, 95–96

literature, illness in (*cont.*)
  Mrs. Dalloway, 91–96
  Pale Horse, Pale Rider, 95
Little Haiti, xvii–xx, xxiii, xxvi, 134
Liu, Joanne, 142
lockdowns, 151, 233
  Covid, 147, 187–89, 192–94, 196–98, 202, 205, 206, 218, 232
  Ebola, in West Point, Liberia, xxvii–xxviii, 146–53, 168–70, 176–81
London, 82, 84
Long Emancipation, The (Walcott), 37
Look Homeward, Angel (Wolfe), 95–96
Loomis, Lafayette, 33
Lorde, Audre, 98–99, 187, 204
Luo people, 44
lymphangitis, 63

M
Maasai people, 44
MacDonald, Ross, 127, 128, 129
MacDougall, Clair, 151–52
malaria, 63, 66, 67, 139, 180
Malcolm X, 36–37
Mallon, Mary, xiv
Manaugh, Geoff, 165
Manchester Evening News, 81
Manigault, Louis, 9–14
  Stafford and, 13–14, 17

marijuana, 224
Marin County Women's Detention Center, 228
Marseille, 214
Marx, Karl, 144
Marxism, 106
Marya, Rupa, xxi–xxii
Mask of Anarchy, The (Ellis), 157–58
Massacre River (Philoctète), 229
maternal mortality, 7
  in slaves, xxviii, 22, 25, 27, 28
Mbembe, Achille, 7, 55
McGovern, Mike, 161
McKay, Claude, 214
McKittrick, Katherine, xviii–xix, 29
Medical Apartheid (Washington), xx–xxi
medical care, *see* health care
"Medical Education in the United States and Canada" (Flexner), 35–36
medical racism, xx–xxi, 37, 68
medical schools, 35–36
  African Americans in, 33–36
  women in, 76–77
Meharry Medical College, 36
Menninger, Karl A., 83
mental health and illness, xv, 83, 100, 148
  influenza and, 83–84
  Woolf and, 85, 86, 88, 92, 102
  prisons and, 136

Mérancourt, Widlore, 221

MERS, 143

Mexico-US border, 232

Meyer, Jaimie, 127, 129

Miami, FL, xix–xx

Lemon City (Little Haiti), xvii–xx, xxiii, xxvi, 134

microbes, microbiology, 46, 49, 56, 82, 145, 165, 168, 231

    bacteria, 4, 42, 50

    germ theory and, 50, 56–57, 79, 84

    human-microbe relationship, xx, 62, 170

    Koch and, 48–52, 56, 62–63, 68

    postulates of, 51

    viruses, 4, 79, 155

    animal-human contact and, 155, 171–72

microplastics, xxii

Miller, William Ian, 174

Misra, Tanvi, 227

Mississippi River, 17–18

mold, 13

monkeys, 171–75

Monrovia, 141, 144–46, 171, 173

More, Ellen S., 77

Mount Pleasant Female Prison, 123–24

Mount Sinai Hospital, 101

Mrs. Dalloway (Woolf), 91–96

Muansa, 44

Mugele, Josh, 141–42

Murdoch, Iris, 186

Muyembe Tamfum, Jean-Jacques, 153–54, 163–64

N

Nagbe, Imanoel, 172

Nama, 68

NAMES Project, 124

Namibia, 57

Narrative of the Life of Frederick Douglass, an American Slave (Douglass), 20–21

Nation, 223, 227

National Health Service, 81

Native Americans, 17–18

nature, 212–17

Nature Medicine, 199

Nazi Germany, 66–68, 72

    concentration camps in, 45, 54

Necropolitics (Mbembe), 55

neoliberalism, 162

New England Female Medical College, 34

New England Journal of Medicine, 142

New Orleans, LA, 1–2

New Orleans Academy of Sciences, 1–2

New York, 120

    Auburn, 119–23, 135

    Bedford Hills, *see* Bedford Hills Correctional Facility Rikers, 126–28, 226

New York (*cont.*)
  prisons in, 114, 136
New York Herald, 80
New York Times, 77, 97, 107, 175
Nice and the Good, The (Murdoch), 186
Niger, 200–201
Nobel Prize, 51–52
Norcom, James, 23–25, 27–30
North Carolina, 22
Nyenswah, Tolbert, 146, 169, 176–78

O
Obama, Barack, 147, 169
Of the Epidemics (Hippocrates), 30
O'Grady, Lorraine, 114
Ohio, 224, 226
"On Being Ill" (Woolf), 73, 91–92
One Long Night (Pitzer), 45–46
On Photography (Sontag), 174
opium, 14, 19
Owens, Deirdre Cooper, 26

P
Pacini, Filippo, 15–16
Pale Horse, Pale Rider (Porter), 95
Pandemic (Shah), 168
Pasteur, Louis, 51
Patel, Raj, xxi–xxii
Pattani, Reena, 166–67

Patterson, Orlando, xxiv
Patterson, William, 6
Pearson, Richard, 78–79
pesticides, xxii
pharmaceutical companies, 179
Philoctète, René, 229
photography, 174
physicians
  African American, 33–36
  Aristotle on, 61, 62
  German, in Africa, 66
  influenza pandemic and, 76, 79
  moral code of, 30
  prisons and, 126–30
  vision of care and, 129–30
  women, 77
Piepzna-Samarasinha, Leah Lakshmi, 100
Piot, Pierre, 153, 154
Pitzer, Andrea, 45–46
plague, 39, 84
  quarantine for, 165
Plague, The (Camus), 159–60
Planter's and Mariner's Medical Companion, The (Ewell), 19
pneumonia, 4
police, 225
  Floyd and, 69, 137, 202, 217
Polk, James K., 15
Pope, John, 18–19
Port-au-Prince, 221, 222, 225
Porter, Katherine Anne, 95
poverty, xxiii, 32, 144

Powell v. Ward, 118

Prateek, Shrivastava, 161

pregnancy, 27

  in enslaved women, 5, 21, 24–28

  maternal mortality and, xxviii, 7, 22, 25, 27, 28

Priest, Chad, 141–42

Primate Visions (Haraway), 216

Prison by Any Other Name (Schenwar and Law), 136

Prison Notebooks (Gramsci), 115

prisons, xv, xxiii, xxix, 100, 105–38, 217, 221, 228

  abolition of, 129, 135–37

  Auburn State, 119–23, 135

  author's visits to, 133

  Bedford Hills, *see* Bedford Hills Correctional Facility

  in California, 135

  Covid in, 220, 226–27

  death accelerated in, 126

  in Florida, 135–36, 217

  Foucault on, 114–115

  Gramsci on, 115–16

  in Haiti, 219–21, 223, 225–29, 231

  HIV/AIDS and, xxviii, 106, 108–15, 116, 119, 125, 127, 132–33, 137

  Huntington Unit, 116

  medical care in, 117–20, 126–30

    private healthcare systems and, 128

  sexual health and, 127

  mental illness and, 136

  migrant centers, xxiii, 133–35

  in New York, 114, 136

  Auburn, 119–23, 135

  Bedford Hills, *see* Bedford Hills Correctional Facility

  Rikers, 126–28, 226

  reform of, 112–13, 117–19, 122, 123, 136–37

  solitary confinement in, 115–18, 120–22, 217

  Tolstoy on, 220

  women in, 119–25, 136

  at Bedford Hills, *see* Bedford Hills Correctional Facility

  health care and, 127

  at Mount Pleasant, 123–24

  organized resistance by, 117

  pregnancy and childbirth in, 122–23

  working class and, 135

  writing and, 131–33

Prostitute Act and Prostitution Protection Act, 189

Proust, Marcel, 60

Q

QAnon, xxvii, 203, 218

quality of life, xxii

quarantine, 152, 164–65, 168, 233

  for bubonic plague, 165

  history of, 165

quarantine (*cont.*)
  for influenza, 77, 82, 84
  for sleeping sickness, 61, 63–64

R
racial apartheid, xxiii
racism, 69, 72, 188–89, 202, 217
  in medicine, xx–xxi, 37, 68
  in science, 2, 67–69
Rafter, Nicole Hahn, 121
Rankine, Claudia, 183–84
Rashid, Ismail, 162
Reconstruction, 32–34
recordkeeping and statistics, 82–83
Reinhart, Eric, 129
Resistance Behind Bars (Law), 117
Resurrection (Tolstoy), 220
Rhodesia, 46
Rikers Island Correctional Facility, 126–28, 226
Ritter, David, 110
Rivera, Aida, 111–12, 114
Robert Koch Institute, 69, 203
Roberts, Dorothy, xx–xxi, 27–28
Roberts, Joseph Jenkins, 157
Robeson, Paul, 105
Rodney, Walter, 62
Romance in Marseilles (McKay), 214
Roy, Arundhati, xxv

Royal Prussian Institute for Infectious Diseases, 51
Russian flu pandemic, 79, 81

S
Saini, Angela, 70
Salk, Jonas, 101
Sambo, Luis G., 143
sanitation, 82
  cholera and, 12, 15
  slaves and, 3, 12
Saurabh, Shrivastava, 161
Savage, George Henry, 83
Sawyer, Samuel Tredwell, 24, 27
Scenes of Subjection (Hartman), 25
Schenwar, Maya, 136
Scherschmidt, Dr., 65
Schilling, Claus, 66–68
Schirati, 64
Schultze, Leonard, 68
science, scientists, 39–40, 48–49, 61, 68, 71–72
  Einstein on, 50
  legacy of, 70–72
  Pasteur on, 51
  racism in, 2, 67–69
self-harm, xxii, 136
sex workers, 231
  in Germany, 184–92
  in United States, 189–90
Shadd, Eunice P., 34
Shah, Sonia, 168
Sharpe, Christina, 221

sickbed, 75, 86, 87, 91, 97, 99–100

sickness, *see* illness

Siegel, Bill, 106

Sierra Leone, 151, 156, 158, 165–69, 172, 178

Sirleaf, Ellen Johnson, 144, 145, 147, 158–59, 163, 169, 177

slaves, xviii, xxiii, xxiv, xxix, xxviii, 1–38

  abolitionists and, 30

  and belief in biological distinction between races, 4–8, 20

  cholera and, 3–7, 9–15, 17, 19, 21

  confinement of, 3–5, 7, 12, 26–27, 38

  emancipation of, 31, 35–37

  escape of, 29, 30

  as extension of enslavers' families, 13

  formerly enslaved people, 31–35

  Liberia as home for, 156–57

  in Georgia, 9

  on Gowrie Plantation, 9–15, 17

  insurance policies on, 10

  literacy prohibited to, 22

  medical care provided to other slaves by, 20–21

  plantation medicine and, 3–9, 17–21

  women, 8

    childbirth and, 4, 8

    Jacobs, 22–31, 37–38

    maternal mortality in, xxviii, 22, 25, 27, 28

    pregnancy in, 5, 21, 24–28

    sexual harassment and assault of, 5, 21, 22, 23, 25–26, 28–30, 37

slave trade, 157, 161

sleeping sickness, 40–49, 52–66, 71

  arsenic-based treatments for, 45, 60–62, 64–65

  medical concentration camps and, xxviii, 45, 54–67

  tsetse flies and, 43, 44, 48, 53, 63–65

smallpox, 32, 179

Snow, John, 16

social distancing, 85, 194, 196, 212, 227

social isolation, xxix

Social Science & Medicine, 172

solitary confinement, 115–18, 120–22, 217

Some Account of the Asiatic Cholera (Cartwright), 3

Sontag, Susan, xi, xxvii, 12, 96–97, 99, 125–26, 174, 231

Spackman, Mary D., 34

Spain, 152

Spanish flu, *see* influenza pandemic of 1918–1919

Ssese Islands, 42, 57

Stafford (enslaved man), 13–14, 17

Stanford Open Policing Project, 225
statistics and recordkeeping, 82–83
Stephen, Julia, 87
storytelling, 130–32
stress, xxi, 218
Students for a Democratic Society, 106, 110
Sudan, 46
suicide, xxii, 83, 136
Superior (Saini), 70
Surowiecki, James, 179
Surviving Autocracy (Gessen), 232
syphilis, 65

T
Tanzania, 42–44, 53, 65
Taylor, Keeanga-Yamahtta, 215
Taylor, Zachary, 15
technologies, fear of, 80
Texas Department of Corrections, 117
Thelus, Theriel, 220
Tilley, Helen, 47
TIME, 145
Times (London), 83, 84
Tocqueville, Alexis de, 120
Tolstoy, Leo, 220
Topol, Eric, 129–30
toxins, xxi–xxii
Trail of Tears, 18
Trouillot, Michel-Rolph, xvi

Trump, Donald, 206, 224, 232
trypanosomiasis, *see* sleeping sickness
tsetse fly, 43, 44, 48, 53, 63–65
tuberculin, 62
tuberculosis, 49, 62, 96, 118, 142, 196–97
Tubman, William, 158
Twilley, Nicola, 165
typhoid fever, xiv–xv, 154
    author's bout with, xi–xv, xxvi, xxvii, xxix
Typhoid Mary, xiv

U
Uganda, 43
Under the Skin (Villarosa), xxi
Undying, The (Boyer), 99
United Kingdom, *see* Britain
United Nations, 222
Human Rights Council, 221
Unruly Women (Faith), 121
Until Proven Safe (Manaugh and Twilley), 165
"Uses of the Erotic, The" (Lorde), 187
Utegi, 65
Utell, Janine, 89
Uyeki, Timothy, 166

V
vaccines, 52
Covid, xxiv, 71, 198–99, 203–5

Ebola, 179
flu, 101
mistrust of, 198–99, 203
van der Groen, Guido, 154
Venters, Homer, 126, 226
Vice Media, 171–75
Villarosa, Linda, xxi
viruses, 4, 79, 155
  animal-human contact and,
    155, 171–72
Voice of the Negro, The (Du Bois),
  139

W
Walcott, Rinaldo, 37
Wallace, Michele, 201
Wang, Jackie, 135–36
Washington, D.C., 32–33
Washington, Harriet A., xx–
  xxi
Weather Underground, 106–7,
  109, 110
Weather Underground, The,
  106
Welch, Rachel, 119–23, 135
Wesseh, Patrick, 149–50
West Africa, 156, 158, 162, 164,
  168, 174, 180, 184
  Ebola outbreak of 2014 in,
    143–44, 146, 152, 164, 166,
    176, 178
  travel restrictions and, 151,
    165–68, 170–71, 176

see also Ebola outbreak in Liberia
Liberia, *see* Liberia
West Point, Liberia, xxvii–xxviii,
  146–53, 168–70, 176–81
white supremacy, 70, 136
Wilhelm II, Emperor, 41–42, 47
Williams, Christiana, 146
Willis, Nathaniel Parker, 30
Wilson, Davidette, 149, 150
Wolf, Silvio, 194
Wolfe, Thomas, 95–96
Woman's Medical College, 76–77
women
  African American, 6, 201
  in prison, 119–25, 135–36
  at Bedford Hills, *see* Bedford
    Hills Correctional Facility
  health care and, 127
  at Mount Pleasant, 123–24
  organized resistance by, 117
  pregnancy and childbirth in,
    122–23
  slaves, 8
  childbirth and, 4, 8
  Jacobs, 22–31, 37–38
  maternal mortality in, xxviii,
    22, 25, 27, 28
  pregnancy in, 5, 21, 24–28
  sexual harassment and assault
    of, 5, 21, 22, 23, 25–26, 28–
    30, 37
Women, Race, and Class (Davis),
  26

Woman's Liberation movement, 119

working class, xxiii, 11, 135, 144

Woolf, Leonard, 91, 102

Woolf, Virginia, xxvii, 85–99, 101–3

   mental health of, 85, 86, 88, 92, 102

   Mrs. Dalloway, 91–96

   "On Being Ill," 73, 91–92

   suicide of, 102–3

World Bank, 149

World Health Organization (WHO), 46, 101, 196

   Ebola and, 141, 143, 152, 160, 162, 166, 167, 178

World War I, 73, 74, 76, 77, 80, 81, 84, 85, 91, 93, 94

writing, 200, 214, 217

   in prison, 130–32

   see also literature, illness in

Wynter, Sylvia, 232–33

Y

Yambuku, 153, 154

Y Care International, 163

Yong, Ed, 177

Yorkshire Evening Post, 79

Z

Ziemann, Hans, 66–68

Zimbabwe, 46

About the Author

Edna Bonhomme is a historian of science, culture writer, and book critic and is a contributing editor for *Frieze Magazine*. She is coeditor of the book *After Sex* and her essays have appeared in *Esquire*, *The Guardian*, *The Atlantic*, *London Review of Books*, *The Nation*, and elsewhere. She earned a PhD in history of science from Princeton University. Edna previously held fellowships at the Max Planck Institute, the Ludwig Maximilian University of Munich, the Camargo Foundation, and Baldwin for the Arts. She has received awards from the Robert Silvers Foundation and the Andy Warhol Foundation. She lives in Berlin, Germany.